ANIMAL HEROES

ANIMAL HEROES

Copyright © Summersdale Publishers Ltd, 2012

Written and researched by Lucy York

Summersdale Publishers Ltd
46 West Street
Chichester
West Sussex
PO19 1RP
UK

www.summersdale.com

Back cover images:
Moko the dolphin courtesy of *The Gisbourne Herald*
Barney the macaw courtesy of Christopher Aloi

Printed and bound by CPI Group (UK) Ltd, Croydon

ISBN: 978-1-84953-207-5

Substantial discounts on bulk quantities of Summersdale books are available to corporations, professional associations and other organisations. For details contact Summersdale Publishers by telephone: +44 (0) 1243 771107, fax: +44 (0) 1243 786300 or email: nicky@summersdale.com.

ANIMAL HEROES

True stories of extraordinary creatures

BEN HOLT

summersdale

CONTENTS

INTRODUCTION

A hero is generally defined as an individual admired for their courage, outstanding achievements or noble character. This anthology gathers together some heart-warming and truly astonishing stories that have appeared in the media around the world, proving that animals can have all three of these qualities in bucketloads, and more besides.

Not only have animals come to the rescue or raised the alarm when someone has been in danger – they have also shown heroism in their hardiness, by triumphing over suffering and surviving against the odds. Some animals have even gone down in the history books as war heroes that have made a significant contribution to the war effort, or as treasured mascots that have provided comfort and boosted the morale of weary troops on the battlefield.

You might expect a pet to come to the aid of its beloved owner in their time of need, and there are plenty of stories in this book concerning animals that have done just that, instinctively acting in the right way at the right time in order to save a life, or they may have provided the love and comfort someone needed to get through a difficult patch. These animals are remarkable for their loyalty and determination, and the stories make for some touching reading.

But then it's rather more surprising to hear of a predatory, wild animal protecting an injured, vulnerable human child. We can never know what motivates an animal normally considered a potential threat to humans to show such compassion. It just proves that even the most dangerous beast can have a softer side.

It's not only humans who have benefited from the benevolence of animal heroes – animals have also lent a helping paw, hoof or flipper to their fellow creatures, protecting members of other species from harm and forming unlikely friendships.

One thing that's certainly striking about this collection of stories is the variety of animals involved: prepare to meet cats, dogs, horses, pigs, dolphins, lions, gorillas, bears, elephants, geese, a tortoise, a beluga and a parrot named Peanut – and that's just to name a few.

OUR LOYAL PROTECTORS

When faced with a perceived threat to their territory, or to a human they view as their fellow family member, some animals will react instinctively to fight off that danger. With their pack mentality, dogs make the perfect guard animals and have long been trained for this purpose by humans. Since these things come naturally to dogs it's not surprising to hear the story of a family pet that has acted to protect its owner from a violent criminal entering their home, for example. But, as the other stories in this chapter show, there are many other animals that will bravely do whatever it takes in order to safeguard a human with whom they have formed a strong bond of friendship, or to ward off an intruder on their turf – often risking their own lives in the process.

OI

One brave dog risked life and limb when an unexpected and shockingly violent attack was made on her family's home…

Patricia Adshead was making a cup of tea when three men burst into her home in Plumstead, south London, brandishing machetes. They threatened to kill her, apparently having mistaken her for someone else. The 62-year-old was shocked and terrified by these violent men, whose faces were hidden by ski masks.

Patricia's ex-husband, who was upstairs and heard the commotion, came rushing down and managed to deter one of the men who was heading up the stairs. But in the ensuing struggle with two of the men his hand was severed and left hanging by a thread.

Patricia was now trapped in the kitchen with her pet dog Oi, a 15-year-old Staffordshire bull terrier. One of the men approached, raising his machete, but just as he went to strike a blow Oi leapt at him and bit his hand. The injured attacker struck out at the dog, leaving a huge gash across the animal's head, but that didn't stop brave Oi chasing him out of the house.

A neighbour raised the alarm and the three men fled. Both Oi and Patricia's ex-husband were rushed from the scene for urgent medical attention. Oi was taken to the

Thamesmead PDSA PetAid Hospital, where she was given 30 stitches.

Oi made a full recovery and lived for another two years, when sadly she died of cancer. Patricia was very sad to be parted from her dear friend, but also very proud to hear Oi would be posthumously awarded the PDSA Gold Medal for her bravery. Looking back on the attack, Patricia said: 'If she hadn't gone for him I would have been dead. She saved my life.'

LURCH

Janice Wolf was very glad to have her bovine friend Lurch around when the young calf spotted a hidden danger in her path…

Janice Wolf was out walking around her animal sanctuary, the Rocky Ridge Refuge in Gassville, Arkansas, one day with her dogs when her 11-month-old African Watusi calf, Lurch, started to act rather oddly. The calf stepped right into Janice's path and turned sideways, blocking her way.

Puzzled by this apparently stubborn behaviour, Janice decided she wasn't taking any nonsense from the calf, and grabbed his horns to push him out of the way. But, as she was about to step forward, Lurch tossed his head, throwing her off balance. That was when she noticed the coiled copperhead snake, lying right where she was about to take her next step. Before she could react, the dogs dashed over to investigate and sadly one of them was bitten. Lurch then trampled the reptile, killing it.

Janice was both astonished by and thankful for Lurch's actions that day. The venom of copperhead snakes, or pit vipers, is not usually fatal to adult humans, although it can cause extreme nausea and damage to muscle and bone tissue in the area near the bite. In Janice's case, however, the effects could have been much more serious, as she was very sensitive to insect bites, and had recently been in hospital with a lung condition, so wasn't in a state of perfect health at the time.

Lurch went on to grow into a boisterous and healthy steer. Watusi cattle are a breed originally from the savannas of Africa and have large, distinctive horns that can grow up to 8 feet from tip to tip. In May 2003, Lurch was awarded a Guinness World Record for the largest horn circumference on a steer at a whopping 37.5 inches.

KERRY

Cows are normally placid creatures, except when defending their young, as one farmer's wife in Scotland discovered…

Fiona Boyd, aged 40 and a mother of two, was at home on her own one day in 2007 at the family farm in Chapmanton, near Castle Douglas, Kirkcudbrightshire. Hearing the cries of a young calf in distress, she ventured out to investigate and saw the baby had been separated from its mother. She decided it would be best to move both the calf and its mother into a shed where they could be together, and approached the calf first.

But just then, the mother cow responded to the calf's cries and, seeing Fiona next to its young, charged over and knocked her to the ground.

'The first thing I knew I was lying on the ground – I thought I was dead,' Fiona told reporters later. Each time Fiona tried to get up or crawl away the bellowing mother cow would slam into her again, knocking her back down, applying the full weight of its body to her. Fiona felt trapped and terrified, and worried that soon the other cows would join in – as sometimes happens in such situations.

Luckily for Fiona, her 15-year-old pet chestnut mare, Kerry, was grazing in the same field and charged at the attacking cow, kicking out wildly. This drove the cow back, giving Fiona the chance to crawl the 20 feet to the electric fence and under it to safety. Her husband Matt, 44, who had been working in another field, took her straight to hospital and she was fortunate enough to escape with just some cuts and severe bruising.

During calving, cows, which are usually very gentle animals, can become extremely protective, and it was unwise for Fiona to attempt to move the calf on her own. After her frightening experience she certainly wouldn't be doing it again. 'I am in no doubt Kerry saved me,' said Fiona. 'If she hadn't been grazing in the same pasture, I really believe I would have been killed.'

DAISY

This little piggy called Daisy ran headlong into the path of danger to protect a little boy…

Seven-year-old Jordan Jones was playing happily in his front garden in the company of the family's pot-bellied pig, Daisy, blissfully unaware of the drama that was about to unfold around him. The Jones' neighbour in the Las Vegas area owned a pit bull, which was normally kept locked up in an enclosure in their garden. But on this day in October 2004, the dog had somehow got out.

The pit bull entered the garden and charged straight for unsuspecting Jordan, but before it could get to the boy brave Daisy rushed over, placing her 150-pound form in its path. The angry dog clamped its jaws on to the pig's head, causing the poor animal to squeal in pain. Hearing the ruckus, Kim Jones, 45, hurried outside to see her pet pig pinned by the dog and bleeding profusely.

She called out for her husband, who emerged with his handgun. The neighbour, also drawn outside by the chaos, gave his permission for Mr Jones to shoot his dog. The pit bull was killed instantly. Kim said, 'It killed me to see Daisy hurt, but if not for her it's possible my son wouldn't be here today.'

Daisy incurred some pretty serious injuries – one ear was severed and there were bites all over her face and jowls – and a vet was called out immediately. Arriving on the scene, the vet realised he didn't have the special anaesthetic needed to operate on pigs, and so he had to improvise – after downing two cans of beer Daisy was sufficiently numbed for him to tend to her wounds. A few weeks later, the pot-bellied protector had made a full recovery.

Sadly, the Jones family were later to be parted from their beloved pet when they relocated to Pennsylvania and were forced to put Daisy up for adoption, but she found a new loving home with pig rescuer Kimberly Moneymaker in Sacramento, California. Daisy's new owner said: 'She's not just a hero, she's the sweetest, most wonderful girl.'

ARNOLD

Policemen might take being called 'pigs' as a compliment once they hear about the porcine crime-fighter known as Arnold…

Becky Moyer, 54, of Stevens Square, Minneapolis, was rather surprised when her boyfriend presented her with a part-Yorkshire, part-Vietnamese pot-bellied pig as a gift. 'Some people get lingerie,' she said. 'I got a pig.' But she soon came to love her new pet and named him Arnold. He weighed about 10 pounds then, but grew to a hefty 300 pounds. Becky also acquired Axel, two months younger and more than 100 pounds lighter, as a companion pig for Arnold. The pair were often found sleeping with their rear cheeks pressed together in the comfy pig house in Becky's back garden, although in winter they would stay inside the house.

One day in February 2001, Becky came home to find her garage door open. Assuming she'd left it that way by accident, she entered – and found two men inside. One of them pressed something that felt like a gun to her back and demanded her purse. Frightened Becky told them it was inside the house and so in they all went together. Once through the door she shouted Arnold's name, and the pig leapt to its feet and grabbed one of the men by the leg with its jaws. The man cried out in pain and cursed, and the two of them fled, leaving a large amount of the miscreant's blood behind. Axel took a backseat while all this was going on, hiding under a chair and squealing with alarm.

Ever since then, Arnold was known as 'the crime-fighting pig' by local police, who always stopped to pet him whenever they were passing by. Arnold's heroic deed won a Minneapolis Police Department Building Blocks award that year, set up to acknowledge the good work of neighbourhood groups that help to build community, solve problems and work with the police.

All the attention Arnold received did raise a somewhat thorny subject – that pigs and other hoofed animals were not technically allowed to be in the city without a permit. There was talk about making an exception to this law for pot-bellied pigs, which had become popular pets.

But Becky was very proud of Arnold for what he did to help save her that day, and also for the role that he and Axel played in bringing her local community together. There were many immigrant families in the area – Russian, Hispanic, Somali and others – and, although language could be a barrier, the sight of a pig united children of all backgrounds in their excited reactions.

SUNSHINE

A criminal realised he'd picked the wrong apartment to break into when he came face to face with a winged fury…

When hairdresser J. W. Erb came home to find his apartment in Williamsport, Pennsylvania, had been burgled, his first thought was for his pet parrot, Sunshine. He had bought the beautiful blue-and-gold hybrid macaw for $1,000 as a gift to himself on his fortieth birthday five years before, and was extremely fond of him. He was dismayed to find a very dishevelled looking Sunshine in the laundry basket on top of his dog's cage. 'My heart sank,' he said. 'He was missing a lot of feathers, and there was blood all over the apartment.'

The man to blame had managed to get away with a camcorder and $100, but not without sustaining extensive injuries from a frenzied attack by the bird. It was very easy for the police to link him to the crime, given that his face looked like it had been rolled in barbed wire. The criminal was brought into the station on an unrelated charge the same day and was already being held when Erb made the 911 call. When officers confronted him with pictures of Sunshine, he confessed to the break-in.

Although undoubtedly relieved that the perpetrator had been caught thanks to his heroic bird, Erb was more worried about poor Sunshine's welfare. Eleven of his tail feathers had been pulled out during the battle and the bird was severely traumatised, remaining silent for three weeks and becoming very clingy. But happily Sunshine did make a full recovery and was soon back to his funny old habits – singing Cher songs and dancing in the shower. 'We've picked up the pieces and moved on,' said proud owner Erb.

STORMY

While out with their horse in the woods one day, two children came across an unexpected foe…

Before Stormy came to be part of the Leonard family in Sulphur, Calcasieu Parish, Louisiana, she had been found in tragic circumstances. The Calcasieu sheriff's deputies were called out to deal with an abandoned horse and found the 30-year-old mare in an emaciated state. She was with a stallion and her yearling colt, and was eating bark off the trees to stay alive.

Stormy was taken to the local non-profit organisation Steeds of Acceptance & Renewal (SOAR), where riding instructor Heather Dionne recognised she had been well trained and had probably even performed in shows in the past. The mare made good progress and, despite her age, in time she could be saddled up and ridden again.

There was an influx of horses at SOAR at the time and so Stormy was placed with a loving family. The Leonards were looking for a horse for their nine-year-old daughter, Emma, to ride, and so they took her in – much to Emma's excitement. The little girl would rush to finish her homework and chores every day so she could go and see her new friend.

One day after school in September, Emma and her seven-year-old brother Liam headed out for a walk with Stormy. This time Emma put Stormy's bridle on, but no saddle, and Liam, who wasn't as comfortable on a horse, began to walk alongside, pretending to be a soldier in a game the pair had played a few times before. They set off down an unexplored path and Liam trailed behind with his rubber-band shooter, keeping a lookout for baddies.

As they reached the end of the trail, Stormy seemed nervous and started to act up, snorting and dancing around, which was strange for the usually composed horse. Emma's efforts to calm Stormy down failed. Liam had just moved in front of them both when a sudden sound from the woods caught their attention. Emma turned to see a large brown wild pig with huge sharp tusks jutting out of its mouth. The children were terrified, especially Emma, as she had heard stories about territorial wild pigs attacking people. She wanted to get them both out of there, but her little brother had frozen to the spot with fear.

When the angry pig stepped between the two siblings, Stormy made her move. She resolutely trotted past the pig and nudged Liam into the woods with her nose, then turned to face the attacker. A stand-off ensued, which was brought to an abrupt end when Stormy spun around and aimed a sharp kick with her back legs at the pig's snout – which was enough to send it off squealing into the woods. The children's mother said of Stormy's brave actions: 'Emma and Liam made it home safely that day because that old horse loves her kids and protected them. She is our hero.'

GUARD ANIMALS

Dogs have been trained as guard animals by humans for centuries, and whether they are guarding livestock or property, they prove very effective at scaring off intruders, raising the alarm by barking or even attacking to defend their territory. The breed of dog and the training will depend on the required task. If a watchdog is required to bark and warn of an intruder, breeds such as Scottish terriers and miniature schnauzers are excellent. Livestock guardian dogs (not to be confused with herding dogs, such as collies), which can be trained to attack and fend off predators, include Maremma sheepdogs and Spanish mastiffs. Rottweilers and Rhodesian ridgebacks make great general guard animals, while German shepherds are favoured as police dogs. However, there are several other options when it comes to effective guard animals...

- **DONKEYS** – donkeys have excellent hearing, allowing them to detect when danger is nearby. As territorial animals, they will instinctively defend their domain, kicking out at and chasing off intruders, while their loud braying raises the alarm. They can be trained to protect domestic property and are commonly used by farmers to watch over flocks of sheep, as they will attack dogs and coyotes. Seventy per cent of donkeys can be trained to guard a flock in this way.

- **LLAMAS** – you might think of llamas as funny, fuzzy-looking creatures, docile, stubborn and with the ability to projectile vomit on demand. But they are actually incredibly alert and inquisitive, and will defend their territory aggressively if they feel it to be under threat. For this reason they are often used as livestock guard animals, but they can also be trained to protect your home. When threatened, they give off a distinctive alarm call. They may also react by chasing, pawing at, or kicking an intruder, or by herding the flock they protect into a tight group and leading them away from danger. They have even been known to attack and kill dogs and coyotes.

- **GEESE** – geese are fairly territorial animals, and they will attack if they feel their domain is under threat. Their loud honking calls make for an unmistakeable alarm signal and they are most effective as guard animals when kept in flocks.

Did you know...

... that Jacob sheep, a rare piebald, multi-horned breed, can be trained as guard animals to protect livestock? In 2007 *The Sun* reported that a farmer in Gloucestershire had trained a flock of Jacob sheep to attack burglars.

'If having a soul means being able to feel love and loyalty and gratitude, then animals are better off than a lot of humans.'

James Herriot

ALL IN A DAY'S WORK

Animals have helped humans in the world of work in countless ways throughout history, most notably dogs; with their superior sense of hearing and smell they have particularly proved useful as both search and rescue and assistance animals. Horses and ponies have also been relied on for their strength, dependability and alertness to their surroundings in roles such as police mounts and for haulage in mines, for example. In this chapter there are some incredible stories of instances when such trained animals have heroically gone above and beyond their duties.

Studies have shown how interaction with domesticated animals can have real benefits to people's health; for example, by helping to calm them and thus lower their blood pressure, or by raising serotonin levels by bringing light-hearted moments to a withdrawn or depressed patient's day. For this reason, the use of therapy animals (animals trained specifically to bring comfort to the unwell) has increased, and it's not just the domain of regular household pets such as cats and dogs; there are also miniature horses and monkeys, among others, some of which you will meet in this chapter.

REGAL, OLGA AND UPSTART

Three steadfast police horses set a good example to the
people of London when the city was under siege…

For a period during World War Two, when the city was the
target of German air raids, police horses were evacuated
from London for their own safety. When it was decided they
should be brought back, they were a reassuring presence
on the streets and for many they symbolised normality
– the sight of smart, mounted bobbies performing their
rounds helped to bolster the 'business-as-usual' attitude
that Britons adopted to brazen their way through the war.
The animals had to work in difficult conditions, stepping
over jagged metal, shards of glass, nails, shrapnel and
rubble, which would have covered the streets after air
raids, without losing their nerve. Sometimes, when a milk
pony or van horse was panicking, perhaps having been
unnerved by a sudden sound or unfamiliar smell from a
recently bombed site, police would bring their own steeds
alongside to soothe them with their composed demeanour.

One police horse honoured for his composure in the
face of danger was Regal, whose stable in Muswell Hill,
north London, was bombed during air raids. On the first
occasion the stables caught fire, yet, in spite of the flames

drawing nearer and the billowing smoke and alarming noise all around, he remained calm, then allowed himself to be quietly led to safety, without any fuss, when help came. His composure meant his rescuers could complete their task without hindrance. Three years later, a bomb fell within yards of the same building, this time covering the horse with rubble, and he incurred some injuries from the flying fragments. Once again, he showed no signs of panic, and it was clear later that his lovely temperament was in no way affected by the traumatic events he had lived through.

Another was Olga, a bay mare that was out on patrol in Tooting, south London, when a bomb exploded. A whole row of houses was destroyed and a huge amount of debris was sent flying in all directions, including a plate-glass window that smashed directly in front of the horse. Terrified, she bolted to a safe distance of about 100 yards, but soon regained her composure. Her rider was able to direct her back to the scene so that they could assist in the rescue operation by controlling traffic.

Then there was Upstart. This horse was on patrol duty in Bethnal Green, east London, when a flying bomb exploded just 75 yards away, showering both him and his rider with debris and broken glass. However, Upstart was unperturbed and continued to work calmly with his rider, controlling traffic around the accident site as if nothing had happened.

In April 1947 all three horses were applauded for their bravery and awarded the PDSA Dickin Medal.

Police horses today

In the UK, mounted police are used primarily for ceremonial, police visibility and crowd control purposes, and escorting the military. In a riot situation, a horse can be used to intimidate and disperse crowds, and the added height can allow the mounted policeman to pinpoint and pull out ringleaders from a crowd. They are often seen at football matches and patrolling the streets at night in busy city centres, as well as in other situations where drunken and violent behaviour can be a problem, where they act as a deterrent.

A half or three-quarter bred horse is best suited to police work, combining the spirit of a thoroughbred with the strength and stability of a draught horse. Horses must undergo six months' extensive training, during which they are taught various skills, including to stand calmly and quietly while a mount answers enquiries from the public, to ride through busy traffic, to move laterally through crowds and to be ridden in the dark. They are also acclimatised to recordings of noises such as military bands, trains and crowds in a controlled environment, before being taken out and exposed to different scenarios. Training is based on a system of encouragement and reward to ensure they become happy, well-rounded, obedient animals.

BEN

A perceptive pony saved his driver from an unseen danger deep underground in a mine...

From the second half of the eighteenth century, ponies were used in mines in Britain for dragging loads of coal from the coalface to the pithead. At the peak of their use in 1913, there were 70,000 ponies underground, until mechanical haulage was introduced on the main underground roads and ponies were confined to the shorter runs. Even as late as 1984, 55 ponies were still in use by the National Coal Board.

The animals were normally stabled underground, fed on a diet of mainly chopped hay and maize, and were only brought to the surface during the colliery's annual holiday. They would work an eight-hour shift daily, hauling up to 30 tons of coal in tubs on narrow-gauge railways – it was a tough life.

As well as providing assistance and company to the miners underground, there were reported instances of ponies providing early warnings of impending danger. One such example involved a handsome chestnut pony named Ben. His driver was in the habit of sharing his sandwiches with the animal when they broke for lunch. One day, Ben kept his distance at lunchtime and nothing could encourage him to come closer – not even an extra large helping of sandwich. As his driver continued to tuck into his lunch, Ben became increasingly distressed, pawing the ground, shaking his head, then prancing around in a circle and whinnying. Eventually, worried by the normally calm animal's agitated behaviour, the driver set down his lunch and followed the pony as it backed away. Just moments later the roof above where he had been sitting caved in and his sandwich box was crushed under falling rocks. The driver had his pony to thank for saving his life.

ROSELLE

Michael Hingson was inside the World Trade Center when the terrorist attacks happened in 2001, an event that must have been all the more terrifying and confusing for him because he was blind…

Michael Hingson relied on his guide dog, Roselle, to lead him as he went about his day-to-day activities – whether he was at work, out shopping or navigating public transport, yellow Labrador Roselle would always be at his side, listening carefully to his instructions and vigilantly leading him around obstacles. They were the perfect team and a strong bond of friendship formed between them.

On 11 September 2001, Michael was inside the World Trade Center in New York City when two planes were flown into the twin towers by terrorists. On that day he was more thankful than ever to have Roselle to assist him. In spite of the terrifying noises as the building buckled under the strain of the collision, the smell of smoke, and the shouts and cries of frightened office workers as they tried to get out of the building in the confusion that followed, the brave dog remained calm and attentive to her work. She led Michael down a total of 78 floors, constantly reassuring him along the way, and then out of the building. But her work didn't end there – she still had to get Michael safely home.

When they were about two blocks away from the building, the first tower began to collapse. Amid the chaos, Roselle stayed calm as they ran for the shelter of the subway. As they re-emerged, the second tower collapsed, covering them with ash. But Roselle still remained composed, and guided Michael to the home of one of his friends, where he was able to wait in safety until the trains were back in action, before travelling home to his worried wife. Many other people were not as fortunate as Michael and were trapped inside when the towers collapsed. Roselle's concentration and refusal to dawdle proved vitally important.

Since then, Michael has become a motivational speaker, drawing on his experience with Roselle that day when he speaks to audiences about trust and teamwork. On 5 March 2002, Roselle and Salty, another Labrador guide dog that guided his owner safely out of the World Trade Center, were both awarded the PDSA Dickin Medal:

'For remaining loyally at the side of their blind owners, courageously leading them down more than 70 floors of the World Trade Center and to a place of safety following the terrorist attack on New York on 11 September 2001.'

DYLAN

In 1999, a search and rescue dog saved lives on two separate occasions…

In March 1999, dog handler Neil Powell of the Northern Ireland Search & Rescue Dog Association (SARDA) was called out to the Mourne Mountains, where four Duke of Edinburgh students had been missing for several hours. With him was search and rescue dog Dylan.

The team were up against exceptionally poor weather conditions, but after a long and difficult search Dylan tracked down the frightened students. They were stranded on a ledge 250 feet above the ground. Although his task was complete, he remained patiently and attentively on duty until the rescue team arrived and lifted everyone to safety.

Dylan and handler Neil were praised by Robert Chambers, the teacher in charge of the students, for their efforts: 'I have a huge admiration for Dylan. It was a full-scale rescue; I didn't know where the students were. The weather was bad and conditions horrendous.'

It was only a few months later that Dylan was to come up trumps once again. In November 1999, the UK Fire Service Search & Rescue Team (UKFSSRT) and the International Rescue Corps were sent to the city of Düzce in Turkey to help with the rescue operation in the aftermath of the 7.2 magnitude earthquake that had devastated the area, causing buildings to collapse and claiming 894 lives. Earlier that year another quake had hit Izmit, 62 miles to the west, killing 17,000, so it had been a difficult year for the Turkish emergency services. Dylan, who was assisting the UKFSSRT, worked tirelessly, crawling between floors of buildings, climbing ladders and leaping across dangerous voids in the search for missing people buried in the rubble. He successfully located two people who had been trapped alive.

On 27 June 2006 Dylan and his brother Cracker were awarded a PDSA Gold Medal: 'For displaying outstanding gallantry and devotion to duty while carrying out official duties.' Cracker was also part of the 1999 Turkish earthquake search team. His particular talent for locating the deceased meant many families were at least able to pay their last respects to lost loved ones, bringing them closure and peace of mind.

The search is on

When a search team have identified a probable location for a subject, air-scenting dogs such as Dylan can be deployed to that area and used to quickly pinpoint them. The dogs can cover large areas of ground in a short space of time, making the work of search and rescue teams much more efficient, and buying valuable time for any injured missing persons in need of urgent medical care. Once they have found the target, these dogs are trained to return to their handler and bark, before leading their handler back to the injured person. Most breeds of dogs can be trained to track in this way but, because search and rescue dogs have to be tough to take part in difficult and often dangerous operations in a range of conditions, larger, physically strong and agile breeds are most commonly used, such as German shepherds, golden retrievers, Labradors and Belgian Malinois. The dogs also need to have good concentration and not get distracted by any wildlife in the area they are searching.

RUSIK

Most cats have a nose for fish, but Rusik's exceptional ability to track down illegally poached sturgeon got him a job with the Russian police…

Siamese cat Rusik first became known to the Russian police force when he wandered into one of their checkpoints in the Stavropol region, bordering the Caspian Sea. The staff couldn't resist feeding the stray kitten with scraps of sturgeon that had been confiscated from smugglers. The Caspian region produces about 95 per cent of the world's caviar and is plagued by smugglers, who come to the area to capture sturgeon. They then sell the sought-after roe in Moscow and other cities for a huge profit. This is a big concern for the Russian authorities as the intensive smuggling could drive the Caspian sturgeon population to extinction.

Rusik was soon adopted as the checkpoint's cat and gained a sharp sense of the taste and scent of sturgeon, which is why the police had the idea of using the keen-nosed cat to help them uncover illegal stashes of the valuable fish.

Rusik proved to have a real flair for his new job – in fact, he was so good at alerting the police to stashes of sturgeon hidden in trucks and other vehicles that he took over the responsibilities of the local sniffer dog. However obscure the smugglers' hiding place was, Rusik was able to sniff it out.

It's not surprising that Rusik was so good at his job – a cat's sense of smell is more sensitive than a dog's. However, cats rarely have successful careers as sniffer animals because it is very difficult to train a cat and get it to cooperate in the same way as a dog.

Rusik's career came to a sad and sudden end when he was hit by a car driven by smugglers – police didn't rule out the possibility that it was a deliberate hit and run. He would be fondly remembered by the Russian police he had helped on so many occasions.

MAGIC

A miniature horse worked her magic on a withdrawn patient and earned herself the title of hero…

Florida-based Gentle Carousel Miniature Therapy Horses is one of the world's only organisations that breeds and trains beautiful miniature horses for the sole purpose of providing therapy and comfort to a range of people in need, including young cancer patients, elderly hospice residents and abused children, whether in group homes, hospitals or hospice-care facilities.

The tiny horses, which grow no bigger than 38 inches high at the withers, are trained from when they are foals to become accustomed to walking up stairs, riding in elevators, and being around wheelchairs and other hospital equipment. First and foremost, however, they are brought up to quietly and calmly accept love and affection from the people they visit.

Magic, a four-year-old dark mare with a white blaze and distinctive blue eyes, had already brought comfort and relief into many patients' lives, but she really made a difference to one in particular. An elderly lady named Kathleen Loper had lived in an assisted-living facility for three years and never spoken a word to anyone during her time there. But when owner Jorge Garcia-Bengochea led Magic into the room, the moment she laid eyes on the horse, Kathleen said, 'Isn't she beautiful?' Hearing those first words from Kathleen was an emotional moment for all the staff, who were thrilled that she became communicative with them from that point onward.

In 2010 the *American Association of Retired Persons* (AARP) magazine named her Most Heroic Pet. Ann Marie Malave, regional activity coordinator and community liaison for the centre, explained that the little horses acted as a 'bridge'.

'They connect to our residents with maybe a memory, a feeling, something that they've done in their lives. It's just such an experience. There are no words to describe it.'

Pets as therapy

In the UK, the charity Pets as Therapy (PAT) trains up therapy animals, and takes them to visit patients in hospitals and care homes around the country. Registered volunteers accompany the animals on their visits, which aim to comfort and provide companionship to people of all ages. There are around 4,500 PAT dogs and 108 PAT cats in the UK, visiting more than 130,000 people every week.

RICHARD

An estate agent from Missouri found an alternative therapy to help her cope with her illness...

Debby Rose, mother of six, suffered from a severe type of anxiety disorder which caused her to have high blood pressure, a racing heartbeat and extreme panic attacks. But rather than seeking to be put on medication, she managed her symptoms through a different type of therapy, delivered by her monkey, Richard.

Debby was the founder of Wild Things Exotic Animal Orphanage, home to 20 monkeys all under her care – and now one was caring for her. With the monkey by her side,

Debby found she was practically free from her symptoms: 'He's an emotional support,' she explained. 'He calms me down and lowers my blood pressure, from his soothing and his eye contact.'

Debby's doctor Larry Halverson approved of the role Richard played in her life, and believed it was a much healthier way of dealing with her problems than resorting to medication. However, not all of Debby's neighbours were as supportive. Richard followed her wherever she went, and people said they felt uncomfortable when he sat with her through hair and nail appointments. Some even filed complaints and, as a result, Richard was no longer allowed to accompany her into restaurants and supermarkets.

Debby later filed a discrimination case against the Springfield-Greene County Health Department and the local businesses that had banned Richard from entering. The case was rejected on the grounds that Richard did not qualify as a service animal. In the US, the Americans with Disabilities Act defines a service animal as 'any dog that is individually trained to do work or perform tasks for the benefit of an individual with a disability, including a physical, sensory, psychiatric, intellectual, or other mental disability'. Other species of animals, whether trained or untrained, are not considered service animals according to this definition. In many other countries around the world, including the UK, assistance dogs are allowed to accompany their owners into places such as restaurants, where they would not usually be permitted, but this does not apply to other animals.

MEET THE WORKFORCE

Creatures great and small, from husky dogs to elephants and ponies to camels, have been tamed and trained by humans to do a variety of tasks for thousands of years, right back to when they first carried heavy loads, used their muscle to help with farming, and hauled primitive forms of transport. Canaries have been used to warn miners of dangerous gases, cats have kept rodent populations under control on ships and in factories, and dogs have been used by hunters to locate and retrieve fallen prey. Today, mechanisation and modern transport have replaced many working animals, but they still play their part in a large number of roles...

- **SNIFFER DOGS** have been used to track down contraband materials such as pirated DVDs, particularly in Asia where piracy is a huge problem.

- **DOGS** are not the only animals used to sniff out drugs – pigs have proved to have an aptitude for it, too, and of course they are commonly used to locate truffles.

- **DOLPHINS AND RATS** have both been used to detect mines – rats can sniff out landmines, and are lighter and less likely to accidently set them off than dogs, while dolphins can attach markers to underwater mines.

- **IN EASTERN COUNTRIES,** such as China and Japan, cormorants have traditionally been used by fisherman to catch fish. The cormorant can eat any smaller fish it catches, but a tie around the base of its throat prevents it from swallowing the larger ones, which are retrieved by the fisherman.

Did you know...

... that a calico cat named Tama held the title of Station Master at Kishi Station in Kinokawa, Wakayama, Japan?

'The bee is more honoured than other animals, not because she labours, but because she labours for others.'

Saint John Chrysostom

WAR HEROES

Throughout history, animals have accompanied man onto the battlefield. Hannibal famously used elephants in his military campaign against the Romans, and Alexander the Great rode his great stallion Bucephalus across Europe and into Asia, conquering people after people as he went. Horses have long served on the front line in the cavalry, while camels, elephants, donkeys and mules have all played their part in carrying heavy supplies and ammunition through difficult terrain, and many cats have served as resident rat-catchers aboard navy vessels. World Wars One and Two saw animals of many species enlisted to help with the war effort on both sides in increasingly inventive ways – dogs were parachuted into enemy territory as spies, messenger pigeons carried vital intelligence and the very first search and rescue dogs worked in the chaos of the London Blitz. In modern times dogs have proved indispensable in sniffing out terrorists' explosive devices, while dolphins have been trained to detect underwater mines. But these useful practical purposes aside, perhaps the most important role animals have played in any war is boosting the morale of soldiers, as lucky mascots, as a source of comfort and reminder of home, and as a reason to keep on living.

VOYTEK

When the 22nd Polish Artillery Transport Company picked up an orphaned bear cub en route to the battlefield, they had no idea of the legendary status their new recruit would one day attain...

During World War Two, Poland was second only to the Soviet Union in loss of life and property per capita. In 1939 it was invaded by both Germany and the USSR, and endured a terrifying war on two fronts, its people terrorised by both the Nazis and the NKVD (the forerunner to the KGB). Any Poles who could, fled, leaving behind their homes and everything they had known and cherished. At the seaport of Pahlevi in Persia a group of them gathered and volunteered for the British forces. They were sent to Palestine to be assigned and were later to become the 22nd Polish Artillery Transport Company.

As the Polish soldiers travelled through northern Persia towards Palestine, one day they came across a poor, starving boy clutching a large bag. When the bag began to wriggle it was opened by Lance Corporal Peter Prendys to reveal a tiny snout and two sparkling eyes. Much to the soldiers' surprise it was an orphaned Syrian bear cub. Prendys picked up the bundle of fur and held it aloft, then cradled it in his arms and fed it bottled milk until it fell asleep. The soldiers, who were delighted with their find, paid the boy for handing the cub over and named him Voytek, meaning 'little one'.

From then on and throughout the war, the bear became an important focus for the affection of the uprooted, homesick Poles. Prendys cared for the cub, feeding him condensed milk and giving him an old washing-up bowl to sleep in. He was an intelligent and inquisitive creature, shy at first, but soon venturing out of his master's tent on his own to explore. The soldiers were worried that when they got to Palestine the commanding officer would not allow them to keep him, but by the time they arrived Voytek's reputation had preceded him, and the by-now-famous bear was given a warm welcome.

Voytek became an integral part of the soldiers' lives: he rode in the front of their jeeps, shared their food, slept in their tents and certainly got up to his fair share of mischief. In 1942, while the company was stationed alongside the Women's Signal Corps in Quisal Rabat, Iraq, he came across the ladies' undergarments hanging out to dry. There ensued a comical scene as the bear trailed the underwear through the camp, much to the embarrassment of the female officers. When reprimanded later, he seemed so miserable that eventually the girls relented and fed him sweets to cheer him up.

In the summer he learnt how to work the shower, but used it so much the hut had to be locked to prevent him exhausting the water supply. One day he was thrilled to find the door left ajar – inside he discovered a cowering Arab spy who had sneaked into the camp to do a recce for a raiding party that intended to steal all the company's weapons and ammunition. As a result, the Arab confessed all and the raiding party were rounded up. Voytek was given two bottles of beer and allowed to splash around in the bath hut all morning as a reward.

Voytek supplied many moments of light relief for the soldiers, but in time he was to see his fair share of heavy action in the bitter war that raged across the continent.

In 1944, the company sailed aboard the *Batory* to Italy, where they would be responsible for supplying all the British and Polish front-line troops with ammunition, artillery shells and food during the big push to take the hill of Monte Cassino, one of the chief obstacles in the allies' quest to reclaim Rome after the fall of Mussolini. They would be thrust into the eye of the storm, constantly exposed to danger from sniper and aerial attack and surrounded by explosions – and Prendys feared for Voytek. He wanted to keep the bear away from the front line, but he was so unhappy at being left behind at camp that they let him ride in the truck with them as they made deliveries of supplies. The sight of his huge head hanging out of the truck window soon became familiar among the military stationed in the area. Initially, the poor gentle animal was terrified by the noises of battle, but this brave bear soon took it all in his stride, remaining alongside his comrades as they worked in the precarious conditions.

One day the men were unloading 25-pound artillery shells and gingerly passing them to each other when Voytek solemnly put out his paws and took one of the shells, cradling it in his furry arms. The shell was live, but Prendys reassured the others that Voytek wouldn't drop it. From then on the bear pitched in and helped the men to unload the heavy supplies. Within a week, they had delivered 17,320 tons of ammunition, 1,200 tons of fuel and 1,116 tons of food to the front line. The company's regimental badge was later redesigned to show the iconic image of an upright bear lifting a heavy artillery shell, in honour of Voytek.

After the war had ended, in September 1945, the company sailed for Glasgow. On arrival Voytek marched with the other soldiers – to the delight of the cheering crowds on the streets of Glasgow. But sadly, he couldn't stay with the army now the war was over and, when Prendys was demobbed in 1947, it was decided Voytek should live out his days at Edinburgh Zoo. The director of the zoo later said that when he watched the bear enter his cage for the first time: 'I never felt so sorry to see an animal who had enjoyed so much freedom, confined to a cage.' He died in 1963, aged 22.

The moving story of Voytek is told in *Soldier Bear* by W. A. Lasocki and Geoffrey Morgan, and the bear's fame endures to this day: he has featured in many exhibitions and there are plaques commemorating him in Edinburgh Zoo, the Imperial War Museum, the Canada War Museum in Ottawa, and a sculpture of him by David Harding sits in the Sikorski Museum in London.

JACOB

In the dead of winter in nineteenth-century Canada, a local farmyard goose surprised the Coldstream Guards with his fearlessness and sense of duty…

In 1838, the British were trying to suppress French Canadian rebels in Quebec and the 2nd Battalion of the Coldstream Guards was sent to help. One evening when Guardsman John Kemp was on sentry duty a large goose walked past and began feeding nearby. As he watched the goose, Kemp noticed it become suddenly tense and freeze. It soon became clear why – a stealthy fox appeared, its predatory gaze fixed on the plump bird. In a state of panic, the goose dashed to hide between Kemp's legs. The guard realised he couldn't shoot the fox, as the noise would alarm the sleeping battalion – they would think the rebels were attacking. Instead, as the fox made a beeline for its prey he swiftly stabbed it with his bayonet. He had saved the goose's life and, in gratitude, it affectionately rubbed its head against his legs.

The goose had come from a neighbouring farm and returned there later that night, but was soon back at the post and regularly attended sentry duty with the guardsmen. They christened him Jacob, and he became a familiar sight and source of company for whoever was on sentry duty.

On one freezing-cold November night, Jacob was to return the favour to the very guardsman who had saved his life. He began to behave unusually: he did not return to the farm as was his habit and instead stayed on duty with the guard. He appeared unsettled, and kept craning his neck to look over towards the farm. All of a sudden, rebels armed with knives charged out of the darkness and headed straight for the guardsman. The heavy snow on the ground had muffled the sound of their approach, allowing them to sneak up on the guard. Jacob instantly rushed towards them, wings flapping, squawking loudly. The diversion this created allowed the guard enough time to fire a shot, rousing the other members of the battalion, and together they fought off the rebels.

Hearing of Jacob's heroic deed, the commanding officer bought the goose from the farmer and awarded him a gorget, a golden collar which he proudly wore as he waddled around the barracks. On returning to England in 1842, the Coldstream Guards took their mascot with them, where he became the centre of attention of the crowds that came to see the famous sentry goose on duty outside the barracks. He is also said to have impressed the Duke of Wellington, who admired the fine bird's devotion to duty. Jacob lived until 1846, and when he died his body was preserved. He can be seen on display at the Guards Museum on Birdcage Walk, London, still wearing his famous golden collar.

MURPHY

A devoted donkey stuck by his master through thick and thin on the battlefields at Gallipoli during World War One, saving the lives of hundreds of wounded soldiers…

The Battle of Gallipoli was one of the greatest disasters of World War One for the Allied forces, when they attempted to knock Turkey out of the war in order to retain control of the Suez Canal and thus a sea passage to Russia. But the Turks turned out to be a greater adversary than predicted, and it became a bitter, entrenched battle that raged from April 1915 until January 1916, when the Allies were forced to admit defeat and withdraw, at the cost of more than 200,000 lives on each side.

During this battle, donkeys were shipped in to carry water supplies and ammunition to the front line, their sure-footedness being indispensible on the rocky, difficult terrain. It's not known exactly where the donkey dubbed variously Abdul, Duffy and Murphy came from, but what is certain is that he formed a deep bond of friendship with John Simpson of the 3rd Field Ambulance of the Australian Army Medical Corps. This Englishman had travelled to Australia at the age of 17 in 1907, and was sent to Egypt, where Australian and New Zealand (or Anzac) troops were being held in anticipation of a campaign against

Turkey. And so, in 1915 he found himself boarding a ship bound for Gallipoli.

On the front line lay Shrapnel Gully, a deep, winding ravine into which the men were packed tightly in hot, dusty conditions and constantly bombarded. It seemed impossible to get any injured men out of there and to the field ambulance, as it would involve moving slowly and being exposed to incoming enemy fire. So they stayed put, until Simpson took it upon himself to tie a red cross to Murphy's brow and lead him down the gully to retrieve them. There had already been large casualties among stretcher-bearers and huge loss of medical equipment, but this didn't dissuade Simpson from his aim. He would tend to injured soldiers in the gully as best he could, and any with the strength to hold on he would lift onto Murphy's back so the donkey could carry them to the safety of the ambulance stations. A record was not kept of how many times they did this, but they surely saved hundreds of lives.

They continued in this way for the next 25 days, and Murphy never faltered in his trust and devotion to his master. Murphy would always keep a slow, measured pace, as if he knew that any sudden movements might cause further pain to his injured charges. The soldiers also believed Murphy could predict incoming shells, his twitching ears providing a warning. It seemed Simpson and Murphy had somehow been given a free pass; for all the many journeys they made up and down, they escaped uninjured. The field hospital commander in charge of Simpson had noted his work, which he carried out of his own initiative, and acknowledged the important boost it gave to the troops' morale.

Then, on 19 May 1915 the Turks launched a heavy counter-offensive, with around 3,000 lives lost on both sides. An armistice was called to give each side time to clear the dead, and Simpson and Murphy would have to work extra hard that day. On leaving camp without having breakfast, Simpson is reported to have said: 'Never mind, get me a good dinner when I come back.'

He took an extra donkey along that day, tied behind Murphy, and as they set off with two injured soldiers they came under fire. Accounts differ on what happened next – some say that both donkeys were killed, others that the injured soldiers were killed and the donkeys delivered the bodies back to base, while others credit Murphy with doggedly continuing on with his wounded charge and saving that soldier's life. But they all agree on one fact: that Simpson was tragically shot and killed. It is not known what happened to Murphy after this sad event, although there is some evidence that he may have later been evacuated with the 6th Indian Mountain Battery.

General Monash highly commended the courageous efforts of Private Simpson and Murphy, and in Australia a bronze statuette of a man leading a donkey with a wounded soldier on its back stands mounted on a marble plinth in the Gallipoli Galleries of Canberra's War Memorial, commemorating the actions of the brave pair.

BANDOOLA

In the war-torn jungles of Burma, a valiant Asian elephant led a refugee convoy safely away from the Japanese in a hurried evacuation…

Born in November 1897, Bandoola was named after General Maha Bandula, a courageous Burmese commander-in-chief who fought for the independence of his country. A man named Po Toke became Bandoola's 'oozie' (elephant driver), training, riding and caring for the elephant right through his working life. He is credited with being the first oozie to train an elephant with kindness rather than by breaking its spirit, and Bandoola is believed to have been the first Burmese work elephant reared from birth in captivity.

As a young elephant he always seemed to be finding himself in scrapes (once he even stuck his trunk into a pot of hot oil), but when he came of age and was put to work, Bandoola showed himself to be an exceptional animal and was soon breaking records. In just one season, he extracted and transported 300 tons of teak an average distance of 2 miles from the forest to the river.

As with all Burmese work elephants, Bandoola was enlisted to assist with the war effort during World War Two and helped build roads and bridges, and carried heavy supplies. In 1944, during the evacuation of the Kabaw Valley in Burma, a party of 45 elephants, eight calves and

198 people were to travel over difficult terrain, climbing up to 5,000 feet and across into Assam, safely away from the Japanese. Po Toke insisted Bandoola should take the lead, as he was the only elephant that knew to keep his head at height by not looking down, and he was sure-footed enough to choose a safe path.

High up in the mountains they reached a treacherous section of the path, where steep steps just big enough to take an elephant's foot had been hewed into the sandstone rock. With Po Toke sitting on his head, Bandoola carefully negotiated the steps, at times looking as though he was standing on his hind legs, and continued along a narrow path with a sheer drop on one side. The other elephants all followed, and whenever they paused their hind legs shook from the strain of stretching up the steps. The party reached their destination, a tea plantation in Assam, with Bandoola carrying a pannier on his back containing eight children sick with fever.

On arrival, Bandoola was celebrated for his courageous achievement, and all the excitement seemed to go to his head – he broke into a pineapple grove and wolfed down 900 pineapples before he was caught in the act! An unpleasant bout of colic ensued, from which he did recover, and he was undoubtedly forgiven his mischievous behaviour in light of his heroic contribution to the evacuation mission.

Bandoola the elephant's incredible story is recounted in full in the eponymously titled book by J. H. Williams, who was also known as 'Elephant Bill', a British Lieutenant-Colonel and elephant expert in Burma, famous for his work with the Fourteenth Army.

Elephants of war

∘ The Carthaginian general Pyrrhus had a huge army of elephants imported from the Atlas Mountains and used them in his invasion of Italy in 280 BC. The sight of these huge, strange stampeding animals caused many of the Romans to flee.

∘ Hannibal marched across the Alps with 37 elephants to launch an attack on Italy, but only one is supposed to have survived the cold journey. He imported more, but the crafty Romans figured out a way of diverting them, having noticed they would get bored and easily distracted after the initial charge.

∘ Before gunpowder was invented, elephants were used to carry great wooden towers on their backs, on which 30 archers could travel.

∘ The Indian army used elephants as battering rams to smash down the defences of fortified towns.

SADIE

In Afghanistan a Labrador used her nose to save the lives of hundreds of military personnel…

In the aftermath of the terrorist attacks on New York City on 11 September 2001, the united forces of the US and the UK launched an offensive against Afghanistan. The primary aims were to seek out terrorist organisation Al-Qaeda, who they held responsible for the attacks, and to remove the Taliban regime in order to install a democratic government. As the war progressed, the Taliban began to develop and use improvised explosive devices (IEDs) as their principal weapon, and so arms and explosive search dogs became hugely important to the US and UK forces during the conflict. A dog can perform a search eight times faster than a human, eliminating costly delays and ensuring troops can be moved from the vicinity if a bomb is detected.

On 14 November 2005, eight-year-old dog Sadie of the Royal Army Veterinary Corps (RAVC) and her handler, Lance Corporal Yardley, were assigned to search for secondary explosive devices after a suicide bomb attack on NATO's International Security Assistance Force in Kabul, Afghanistan. Leaving a second bomb is a typical terrorist tactic in these situations.

The pair began their search and it wasn't long before Sadie took up the alert stance to indicate she had found something. She had sniffed out a booby-trap bomb concealed in a pressure cooker, hidden behind a 2-foot-thick concrete blast wall within the United Nations compound. It was packed with explosives and rigged up with a remote-controlled detonation device. Had the enemy succeeded in setting off the bomb, it could have killed and injured hundreds of military personnel in the surrounding area. Thanks to the Labrador's razor-sharp senses, personnel could be evacuated to a safe distance and the device was disabled by the bomb disposal unit.

In 2007 Sadie was awarded the PDSA Dickin Medal 'for outstanding gallantry and devotion to duty'. Lance Corporal Karen Yardley attended the ceremony at the Imperial War Museum, along with Sadie, and said: 'She's a lovely dog and I'm very proud of her.'

Dogs like Sadie also play an important role in raising morale among the troops, many of whom miss their own dogs back at home. Handlers do form a strong bond with the dogs they work with, but the animals are given uniform training so they can work with more than one handler. While the dogs usually do two six-month tours, handlers work on rotation and only complete six months in every two and a half years.

Underwater detectives

Dogs like Sadie aren't the only animals to have been set to work seeking out explosives: the US Navy's Marine Mammal Program has trained dolphins to detect underwater mines – a method most famously used during the Iraq War. With their echo location ability, bottlenose dolphins can locate mines with great precision, and they are light and deft enough to be able to mark the mines without triggering them. Sea mines are designed not to detonate when marine life such as sharks, whales and dolphins swim by, and so the chances of a dolphin being hurt when carrying out this work are low. The dolphins are trained using similar methods to those used for police and hunting dogs – and are given rewards such as fish on correct completion of a task.

MARY OF EXETER

During World War Two, a tenacious carrier pigeon
named Mary of Exeter somehow always managed
to deliver her messages safely…

Carrier pigeons had played an important role during the Great War, and yet as World War Two loomed on the horizon the British war office had no plans in place for a messenger-pigeon programme; with inventions such as radar, radio and the telephone to rely on, they thought they could do without them. But they were wrong. When the first British planes were forced down into the sea it soon became apparent that with water-damaged radios there was no way the pilots could communicate their position and request help. An urgent call for pigeons was sent out and private pigeon fanciers offered thousands of birds up for service. The pigeons also went on to do important work on the front line during the war, and some were dropped behind enemy lines in France in the hope members of the resistance would find them and send them back with vital intelligence.

Homing pigeons are raised in lofts and trained to return to them, where they know they will be fed. Wherever they are released, these birds are always able to find their way home. It might be due to an inbuilt compass, or perhaps they recognise landmarks and geographical features – no one really knows how they do it. In war time, messages written in code could be placed inside a tiny canister attached to the bird's leg. These incredibly resilient, strong birds have been know to fly as far as 1,500 miles and to battle against gale-force winds and torrential rain in order to return home – so they really were a reliable form of communication where technological methods had failed.

Mary Stewart was one such carrier pigeon that served under the British National Pigeon Service during World War Two, from 1940. Such was her determination to return home on each mission that she did so despite numerous injuries and terrible ordeals during her five-year military career, by the end of which she had 22 stitches on her poor little body. On one occasion Mary vanished for nearly a week. When she finally arrived home her message was intact, but her neck and right breast had been ripped open, wounds most likely inflicted by one of the hawks that the Germans released in the Pas-de-Calais area to terrorise messenger pigeons.

Carrier pigeons were also constantly targeted by enemy gunmen, and two months after her run-in with the hawk Mary went missing again – this time for three weeks. When she pathetically fluttered back into the loft, there were three bullets in her body and part of her wing had been shot off. As if that wasn't enough to endure, she also suffered an attack on her home. During the German raids on Exeter a large bomb fell outside her loft, killing many of the other pigeons inside. Mary was very distressed by these traumatic events, but after a short break she was back at work. Within ten days she was picked up in a field. She was painfully thin, and had a huge gash on her head and more wounds all over her body. Her owner carefully nursed her back to health but, due to the injuries to her head, she needed to wear a leather collar for support until she was fully recovered.

In November 1945 she was awarded the PDSA Dickin Medal, 'for outstanding endurance on war service in spite of wounds'.

Homing is a natural instinct for pigeons, but it has been noted by handlers that some are more determined and less likely to be discouraged by difficult conditions and obstacles than others, and so the choice of which pigeon to entrust with an important mission had to be made carefully. Mary of Exeter certainly proved to be a reliable choice, time and again.

Winged messengers of war

The Sultan of Baghdad was sending messages out on the wing as far back as 1150 BC, and pigeons were used to spread the word of Ceasar's conquest of Gaul. News of Wellington's victory at Waterloo was also delivered by pigeon, and the birds were instrumental in transporting thousands of letters to the besieged Parisians during the 1870 siege. Although the telegraph and radio had been invented by World War One, they often broke down; more than 100,000 pigeons served Great Britain in that war, 95 per cent of those successfully delivering their messages.

SAM

Sam became the first army dog to win the PDSA Dickin Medal since 1944, for disarming a gunman while on duty in the Balkans…

In the late 1990s, after the Bosnian War of 1992–1995, the Balkans were still a hot-spot for conflict, with repeated uprisings as ethnic tensions played out across the troubled region. In 1998, British troops were stationed at the town of Drvar in Bosnia-Herzegovina, where the return of displaced Bosnian Serb citizens had brought opposition from resident Croats, culminating in a series of riots and murders.

Dog handler Sergeant Iain Carnegie of the Royal Army Veterinary Corps Dog Unit, from Melton Mowbray in Leicestershire, was on duty with his German shepherd, Sam, when a gunman opened fire in the town. Sam wasted no time at all in charging after the man and skilfully bringing him down. 'Sam performed brilliantly – just like a training exercise,' said Sergeant Carnegie, who was able to rush to Sam's side, disarm the man and retrieve the loaded pistol.

Sam also prevented a mob armed with crowbars, clubs and stones from attacking ethnic Serbs in the area six days later. The mob surrounded a group of around fifty Serbs, but the dog held them off until backup arrived.

When Sam was later awarded the PDSA Dickin Medal, Sergeant Carnegie proudly said: 'Sam displayed outstanding courage in the face of the rioters, never did he shy away. I could never have attempted to carry out my duties without Sam. His true valour undoubtedly saved the lives of many servicemen and civilians.'

SIMON

A cat is credited with saving the lives of Royal Navy officers on board the HMS *Amethyst* during the Chinese Civil War in 1949…

In 1948, the British frigate HMS *Amethyst* was stationed in Hong Kong. One day, crew member George Hickinbottom, just 17 at the time, found an undernourished black-and-white cat wandering the dockyards. George smuggled the cat, soon to be named Simon, aboard the *Amethyst*, where he set to work catching and killing rats on the lower decks, and quickly became popular with the crew as the ship's unofficial lucky mascot.

In April 1949, the *Amethyst* was travelling from Shanghai to Nanjing on the Yangtze River to relieve the HMS *Consort* from duty. The *Consort* had been standing as guard ship for the British Embassy there due to the Chinese Civil War between the Kuomintang and the Chinese Communists. As the *Amethyst* travelled upstream, it came under siege from the People's Liberation Army. Later to become known as the Yangtze Incident, the siege lasted for 101 days. Throughout that time, Simon was hard at work killing off rats that had launched their own siege on the ship's food supplies, having gained access to the vessel in great numbers as it was moored up in the river. Running out of food supplies would have been catastrophic for the crew –

there was no way for them to replenish stocks while under siege.

Very early on in the siege, Simon suffered severe shrapnel wounds when an onslaught of shelling penetrated the captain's cabin, killing Commander Bernard Skinner.

Simon managed to drag himself out of the cabin onto the deck, where he was quickly taken to the ship's medical bay. Medical staff treated his burns and removed four pieces of shrapnel from his wounds – they didn't expect him to last the night, but against all odds he made it through and went straight back to his duties of keeping the rat population under control. He also made regular appearances in the sick bay, helping to raise the morale of wounded sailors, many of whom were just teenagers.

By the time the ship escaped from the river, news of Simon's heroic actions had spread and he'd become something of a celebrity. He was showered with praise at every port the ship called at on the passage back to Britain and received a hero's welcome when the *Amethyst* finally returned to dock in Plymouth on 1 November 1949. Letters from fans and admirers arrived in their thousands – there was such a deluge that he was assigned his own naval officer to deal with the fan mail.

Although a celebrated hero, Simon still had to go through quarantine on his arrival back in the UK. Sadly, he died there three weeks later from a complication of the viral infection caused by his earlier wounds. A funeral with full military honours was held for him at the PDSA Animal Cemetery in Ilford, Essex. Hundreds attended, including the entire crew of the HMS *Amethyst*. He was posthumously awarded the PDSA Dickin Medal, the only cat ever to have received it, and was also recognised with the rank of able seaman.

Commander Stuart Hett, speaking at a ceremony in 2007 to commemorate Simon's bravery, said: 'Simon's company and expertise as a rat-catcher were invaluable during the months we were held captive. During a terrifying time, he helped boost the morale of many young sailors, some of whom had seen their friends killed. Simon is still remembered with great affection.' The director general of the PDSA, Marilyn Rydström, commented that: 'The power of animals to support and sustain morale in times of conflict can never be underestimated.'

TIRPITZ

A deadly sea battle during World War One resulted in an unusual new addition to the crew of the HMS *Glasgow*...

On 14 March 1915, the British ships HMS *Glasgow* and HMS *Kent* engaged in battle with the German cruiser SMS *Dresden*, in the waters off Chile near Juan Fernandez Island (today known as Robinson Crusoe Island). Forced to scuttle their own vessel, the German sailors all abandoned

ship, leaving just one crew member standing by the sinking vessel to the bitter end – a pig. The sailors aboard the *Glasgow* watched the *Dresden* disappear beneath the waves and their astonishment grew as the animal bravely struck out through the choppy water.

One sailor decided to rescue the animal. He leapt into the water and swam towards it, nearly drowning when the pig began to panic. But eventually he managed to grab it and the pair were hauled back aboard the *Glasgow*. As the crew gathered around the new arrival on deck, they were unanimous in their decision to make it the ship's mascot and to name it Tirpitz, after the famous German admiral of the same name. They also thought the pig very brave for not abandoning the sinking *Dresden* and so honoured him with their own special award.

Tirpitz quickly adapted to life aboard the *Glasgow* and was a popular member of the crew. There wasn't much he could do to help out with the war effort, but he did play a very important role in helping to boost morale. After many long months at sea in difficult conditions the sailors were war-weary and homesick – the sight of chubby Tirpitz snuffling around the decks on pig business was just what they needed to lift their spirits.

After completing his ship mascot duties, Tirpitz lived out the rest of his days at the Whale Island Zoo, Portsmouth, where many an animal mascot was sent to retire. His life came to a somewhat sad end in 1919 when he was auctioned off for charity – but even in death he proved his worth, as the £1,785 that was paid in exchange for his pork went to the British Red Cross. Tirpitz's head was mounted and put on display in the Imperial War Museum, London, and his story was later featured in the Imperial War Museum's 'The Animals' War' exhibition.

Marvellous mascots

When it comes to military mascots it seems that anything goes. Here's a round-up of some of the more unusual ones:

- Bobby the antelope of the Royal Regiment of Fusiliers.

- Two ferrets named Imphal and Quebec, of the 1st Battalion, the Yorkshire Regiment.

- Honourable Regimental Sergeant Major Nils Olav – this King penguin, who lives at Edinburgh Zoo, was bestowed the honourable title by the Royal Norwegian Guard.

- Jane the cow, who provided Devonshire milk and cream to soldiers during World War One.

- Timothy, a tortoise that lived to be 165 and served as mascot aboard the HMS *Queen* and *Princess Charlotte*.

SMOKE

US Marines stationed in Iraq made a new battle buddy when a donkey wandered out of the desert and into their lives…

In July 2008, US marines were stationed in the Al Anbar province, Iraq, involved in the ongoing Operation Iraqi Freedom. Camp Al Taqaddum, an old Iraqi airbase not far from Fallujah, was on a barren plateau, set against a backdrop of inhospitable desert, and for many US troops it was their first taste of Iraq, with new deployments regularly landing on the dusty desert runway.

Colonel John Folsom had been sent there to take over as camp commandant. One day his commanding general showed him a video on YouTube, taken from a security camera at a Marine Corps base of a group of marines and soldiers in a mad chase, struggling to round up a wild donkey that had got into their camp. Folsom thought the video was hilarious and mentioned in passing to Sergeant Juan Garcia that they should try to capture their own desert donkey should one ever come their way.

One morning a few days later, Folsom woke to a loud hee-hawing sound near his window. He went outside and, sure enough, Garcia had taken him at his word and captured a donkey he had spotted wandering around the camp – which was now tied to a tree, looking rather thin and hungry. In the dry and searing temperatures of the desert there wasn't much grass to be found, and so the soldiers fed the donkey granola bars until some hay could be sourced. They christened the animal Smoke and it was to be the start of a longstanding friendship. 'He was a battle buddy if you want to call him that. You know he did a lot of good for us morale wise,' explained Folsom, who began to go on daily walks with Smoke and had a coral built for the donkey.

Strictly speaking, the marines weren't allowed to keep pets, but after a navy captain who was a psychiatrist picked up on the value of Smoke as a 'therapy animal' he gained official status and was assigned to the 'combat stress' department. Soon marines were sending pictures of Smoke to their families, giving children back home something positive to talk about with their much-missed dads.

At the end of Colonel Folsom's deployment, before journeying back to the US, he arranged for the replacement marines, of Camp Lejune, North Carolina, to continue to look after Smoke in his absence. But when the Campe Lejune marines in turn moved on they left him in the care of a local sheikh, and Folsom later discovered that Smoke was not being cared for properly and was wandering loose around the village.

By now Folsom had formulated a plan for Smoke – if he could get him back to the US he would not only have a loving and safe home, he could also perform an important role as a therapy animal for returning wounded soldiers. But first he had to negotiate with the sheikh for

Smoke's return to the marines, and then jump through the various bureaucratic hoops set by customs, agriculture and airline officials. He enlisted the help of the Society for Prevention of Cruelty to Animals (SPCA), who helped him to successfully arrange for the transport of Smoke from Iraq via Turkey to the US.

People around the world had begun to follow Smoke's story online through the SPCA's website, so he was already famous by the time he arrived at his new home in Omaha, Nebraska, where he would continue to bring smiles to the faces of many soldiers and marines working as a therapy animal with the Wounded Warriors Family Support initiative, founded by Folsom.

JACKIE

Jackie was a beloved family pet that never left his master's side, even when it meant following him to war...

In August 1915, Albert Marr from Villieria, on the outskirts of Pretoria, South Africa, enlisted for service with the 3rd (Transvaal) Regiment of the 1st South African Infantry Brigade. He also brought with him his treasured companion Jackie – a baboon. Although it was somewhat irregular for pets to join their owners on the battlefield, for some reason the army decided to make an exception and approve Albert's request to bring him along. The other soldiers were very impressed by the baboon's excellent behaviour and it wasn't long before he was made the official mascot of the regiment. Once the troops arrived in England he was kitted out with a uniform, including a cap, buttons and regimental badges.

Early in 1916, the 1st SAI Brigade was dispatched to Egypt to take part in the Senussi Campaign. The aim was to push back the Senussi tribe which, under the influence of Germany's ally Turkey, had flooded into the country. At the Battle of Agagia on 26 February, Albert took an enemy bullet in the shoulder. A very distressed Jackie stayed at his master's side, doing what he could to comfort Albert and licking his wound until the stretcher-bearers arrived. After this touching incident, the other soldiers began to view the baboon as a comrade, and he was involved in all parts of military life: he participated in drills and marches, entertained the men with his antics and with his keen eyesight and acute hearing he excelled at night-watch duties.

Stationed at the Western Front, Jackie would have been exposed to some of the worst horrors of World War One and did not escape the war unscathed. In April 1918, the company came under heavy shelling during a retreat and Jackie incurred shrapnel injuries on an arm and leg. The leg later had to be amputated.

When their active service came to an end, Albert and Jackie were much celebrated. As they journeyed home,

Jackie proudly sported his military colours: a gold wound stripe and the three blue service chevrons, one for each year of front-line service. He even received official papers and a military pension when he was discharged at Maitland. At the Peace Parade on Church Square, Pretoria, on 31 July 1920, to officially welcome back the 1st SAI Brigade, Jackie was the centre of attention and was awarded the Pretoria Citizens' Service Medal.

THE PEOPLE'S DISPENSARY FOR SICK ANIMALS

A woman with a vision

The People's Dispensary for Sick Animals (PDSA) was founded by Maria Elisabeth Dickin. When visiting the impoverished in the slums of London's East End to do social work, she was shocked and horrified more by the plight of the animals – emaciated dogs and cats, farm animals huddled in crowded back yards and overworked horses and ponies. The inspirational woman became determined to raise the status of animals in society and improve the standard of their care. By November 1917 she had opened her first People's Dispensary for Sick Animals of the Poor, in Whitechapel. Outside a notice read: 'Bring your sick animals. Do not let them suffer. All animals treated. All treatment free.'

Within six years she had a fleet of horse-drawn mobile clinics treating animals throughout the country, and by 1935 Maria had established five PDSA hospitals, 71 dispensaries and 11 motor caravan dispensaries, as well as others abroad in Egypt and Greece, and later South Africa and Palestine. Today, the PDSA continues to ensure that sick and injured pets of those unable to afford private veterinary fees can still receive free treatment.

'Today we are all thinking about what each of us can do towards making the world a better place for every man, woman and child to live in. We must not forget to include the animals in our programme, they too must have a better world to live in.'

Maria Dickin

Did you know...

... that when the first PDSA dispensary opened in London the local poor people were initially suspicious because it was labelled as 'free'? Once they had overcome their fear, police had to be called in to control the crowds as hundreds flocked to the doors with their sick animals.

We also serve

During World War Two, Maria Dickin CBE was aware of the incredible bravery displayed by animals on active service and the Home Front. Inspired by their devotion to man and duty, she created the Dickin Medal specifically to honour animals in war. The large bronze medallion bears the words 'For Gallantry' and 'WE ALSO SERVE' within a laurel wreath, and hangs from a ribbon striped green, dark brown and pale blue representing the naval, land and air forces.

'The medal is recognised throughout the world as the animals' Victoria Cross and is the highest award any animal can receive for bravery in the line of duty.'

Marilyn Rydström, director general of PDSA

ANIMALS TO THE RESCUE!

Animals have rescued humans from dangerous situations in a surprising number of ways, whether by pushing them out of reach of a fire, pulling them to safety, providing essential warmth and sustenance to someone exposed to the elements or even calling the emergency services. Many of the stories in this chapter feature heart-warming human-animal friendships that have resulted in the animal saving the human's life; with others, the animal just happened to be in the right place at the right time.

JEUNE MARK

When the Black Saturday bushfires swept across Australia's Victoria state, the death toll reached 173 – and Anthony Sexton very narrowly missed being the 174th...

On and around Saturday 7 February 2009, fires raged across the state of Victoria, resulting in Australia's highest ever loss from bushfires – 173 people died and 414 were injured. Preceding the fires came some of the worst weather conditions ever recorded, with temperatures soaring well into the forties. This intense heatwave, combined with nearly two months of little or no rain and wind speeds of more than 62 mph, meant the fires quickly spread over large areas. As many as 400 separate fires were recorded, and the events of that day and its aftermath came to be known as Black Saturday.

During the bushfires, Anthony Sexton thought he was safe in his 130-year-old farmhouse in the foothills of the Strzelecki Ranges. He could see the fire decimating the bushland behind his house, but it seemed to him that the flames were not set to come in his direction. After a sudden drop in temperature, smoke billowed towards the house and Andrew finally accepted it was time to leave.

He led his horse Jeune Mark (the son of 1994 Melbourne Cup winner, Jeune) some distance down the road before the path was blocked by a wall of flames. Thinking they might be able to race back the other way, he turned, only to see the fire pouring down the hill to surround them. Andrew was already thinking to himself: 'This is where I'm going to die.'

But Jeune Mark had other ideas. The horse swiftly nudged Mr Sexton over a barrier rail, sending him rolling down into Traralgon Creek. There he lay for two hours, as the inferno raged, and he could do nothing but listen to the terrifying sounds of trees exploding in the heat of the fire and the wind roaring above.

When the flames had subsided and he judged it was safe for him to emerge, Andrew found everything around him had been razed to the ground. His thoughts turned immediately to Jeune Mark and the fact that he must surely now be dead. And yet the horse *had* survived and was there to meet Andrew when he returned to his house, which was still alight. Courageous Jeune Mark must have had to race through scorching flames and choking smoke to reach his home, but miraculously, other than some burns around his eyes and nose, he had somehow made it through unharmed. Andrew had lost his home to the bushfire, but he was thankful that he hadn't lost his dearest friend.

A close call

In southern China, a 36-year-old man named Fu Min slipped while trying to take a photograph at a race and fell into the path of the oncoming riderless horses. He was on the verge of being trampled when one of the steeds grabbed his arm with its mouth and dragged him out of harm's way.

SHANA

A German shepherd-wolf mix came to the rescue
when its owners were trapped after a violent storm…

Eve and Norman Fertig ran the Enchanted Forest Wildlife Sanctuary in Alden, New York, and cared for many different types of animals there, usually around a dozen at any one time. Eve also taught adults to be wildlife rehabilitators. The couple, who were in their eighties, did their work voluntarily, paying for their own teaching licences and for the upkeep of the animals out of their social security cheques. One such animal they rescued was a sick, two-week-old half-wolf, half-German-shepherd puppy, Shana. Almost seven years later Shana was fully grown, and the loyal dog still followed her owners around wherever they went.

On 12 October 2006 it was a clear evening, and the Fertigs were out feeding and exercising the animals in the forest sanctuary on their property, as was their regular routine. But at around 7 p.m. an unexpected storm set in, causing a power cut in the Fertigs' house. When they saw the lights go out they realised things were bad: they went to investigate and a huge tree came crashing down across their path. More trees had fallen, effectively trapping the elderly couple between the aviary and the hospital building of the sanctuary, and now snow was falling. Soon it would begin piling up into drifts around them.

'I think we could die out here,' Eve said to her husband. There was no way the elderly couple could climb over the trees, for fear of injuring themselves, and as it had been a clear day just hours earlier neither of them were wearing warm outdoor clothes. They huddled together for warmth, but by now it was 9.30 p.m. and temperatures had plummeted.

Help was to come, however, from an unexpected source, when Shana's friendly face greeted the Fertigs. The 160-pound dog had dug a path through the snow and beneath the fallen trees to find her owners. She looked at them and barked, as if telling them to follow her back down the mineshaft-like tunnel. At first Norman was reluctant – he wasn't too fond of enclosed spaces after some unpleasant experiences in a foxhole in Japan during World War Two. But Eve was quite forceful about the matter: 'Norman, if you do not follow me, I will get a divorce.' Evidently horrified by the thought, Norman agreed.

Shana disappeared back down the tunnel until around 11.30 p.m. – presumably during this time she was extending the tunnel as far back as the door of the house. When she came back she grabbed the sleeve of Eve's jacket, draped the 86-pound woman over her back and began to move through the tunnel. Norman grabbed hold of Eve's legs and followed. They made their way slowly down the tunnel, underneath fallen trees and through an opening in the fence to arrive at the house at around 2 a.m. Exhausted, the couple fell through the door and just lay there on the floor, the dog lying on top of them both to keep them warm – as there was no electricity or heat in the house they most likely would have frozen without her.

When concerned neighbours hadn't been able to contact the couple on the phone that night they had called the Town Line Fire Department, and the fire fighters arrived the next morning. They offered to take the Fertigs down to the fire station to shelter along with around a hundred others stranded by the storm, but when they were told they couldn't bring Shana they refused to leave their home. Instead, they sat it out for three days until the station emptied of people and they could go there with their pet. The fire fighters worked at clearing the trees from the grounds, and brought food and water for the couple and their animals. Once again, Shana provided a vital source of warmth for the couple during this time, when there was no electricity, heating or hot water in their home. 'She kept us alive. She really did,' Eve said.

At the fire station, Shana was spoilt rotten by the fire fighters, who had never seen anything like the tunnel she had built. Shana's story gave the exhausted fire crews hope and helped boost their morale during the gruelling work in the aftermath of the storm. Shana was later to receive the Citizens for Humane Animal Treatment's (CHAT) Hero's Award for bravery, which is usually only given to humans. The plaque, complete with Shana's picture, hangs in pride of place in the Fertigs' living room.

COW 569

A New Zealand dairy farmer found herself an unusual
life ring when floods swept through her farm…

The area north of Wellington, New Zealand, was affected
badly when floods hit the country in 2004. A burst of cold
air blowing in from the Antarctic ice shelf combined with
moist air from a weak tropical low in the north, producing
wind and rain on a scale seen only about once every ten
years, with wind speeds peaking at 104 mph. Hundreds of
North Islanders were forced to evacuate their homes and
insurers estimated the cost of damage at £40 million.

Kim Riley, 43, was out early in the morning on her dairy
farm in Woodville, trying to head off half her herd, which
were moving in the direction of the floodwaters, when she was
swept away by the current herself. Weighed down by her wet-
weather gear, she struggled in the torrent, as assorted debris
floated past her and panicked members of the herd swam over
her at speed. She received a few hefty kicks as they passed.

'I drank a fair bit of water and it was foul muck. There
was lots of flotsam around and weeds get tangled up in
your arms: it was quite exhausting,' she later told reporters.
Desperate, she reached out to clutch at anything she could
– a tree, the top of a fence – but missed both as the current
carried her ever onwards. She was terrified she would be
swept out along with the floodwaters into the river itself,
and then she would have no hope of surviving.

Luckily for Kim, serendipity threw her a life ring – in the form of a floating Friesian. Looking back she saw one of the last of the herd bearing down on her. This time, instead of trying to avoid the beast, she threw her arm over its neck and grabbed on tight to its mane. The warmth of its body was reassuring and her panic began to subside. 'Take me on home,' she said to the cow.

When the pair finally reached solid ground they both slumped, catching their breath and trembling. There was more good news to come – most of the herd had been saved, having managed to gather on higher ground. Only 15 of her 350-strong herd had been lost to the flood, and Cow 569, her saviour, was due some special attention and recognition, as straight-talking Kim unsentimentally put it: 'She's an old cow, an ugly old tart, but I'll have to say "thank you" to her.'

TOMMY

Police in Columbus, Ohio, responded to an emergency call from Gary Rosheisen's address, but were surprised by what they found waiting for them…

Gary Rosheisen was wheelchair-bound and suffered from osteoporosis, as well as occasional mini-strokes that affected his balance. He lived alone in an apartment with his orange-and-tan striped cat, Tommy, who he had got three years previously. The cat was a source of company and comfort to Gary, and seemed to have a calming effect on him, thereby helping to lower his blood pressure.

One Thursday night, Gary fell from his wheelchair and couldn't get himself back up off the floor. He was equipped with a medical alert necklace to contact paramedics in such an emergency, but on this occasion he wasn't wearing it, and there was no way he could reach the emergency cord above his pillow. He was stuck.

Shortly after the time of his fall, the police received a 911 call which they traced back to Gary's apartment – but they couldn't hear anyone on the end of the line. Thinking it might be a misdial, the police called back to check everything was OK. When no one picked up they prepared to go to the apartment and check on the situation.

Arriving at Gary's home, the police were greeted with the sight of a cat lying by the telephone on the living room floor.

Then they saw Gary, who was on the ground near his bed. Gary explained to the police that Tommy must have made the 911 call. Officer Patrick Daugherty, who was at the scene, said, 'I know it sounds kind of weird.' But there wasn't any other possible explanation.

Gary had tried to teach Tommy how to dial 911, but until that day he wasn't sure that his training had sunk in.

Gary kept his phone on the living room floor – it only had 12 small buttons, including a speed dial for 911 just above the speaker-phone button. So it would have been fairly easy for Tommy to press the right button with his paw when he saw his owner was in trouble. 'He's my hero,' Gary said.

MANDY

When an elderly man was injured while out alone on his farm in a remote area, he knew he had a fight for survival on his hands...

On 8 October 2002, Noel Osborne, 78, was at work out on his farm in a remote area near Benalla, Australia, when a cow knocked him over. He fell hard and his hip was broken, leaving him unable to move. He called out for help but there was no one for miles around: he was alone.

For several frightening days and nights Noel was stuck outside in an open field, exposed to the elements. The days were hot but the nights were bitterly cold, and at certain times storms broke overhead. Noel would never have survived the terrible ordeal if it wasn't for his pet nanny goat, Mandy, and his loyal border collie, also named Mandy, strangely enough.

On the first evening the nanny goat turned up and approached him. He managed to reach for an old bottle and Mandy allowed him to milk her. The milk provided vital sustenance, and during the night both the goat and the dog snuggled down either side of him to keep him warm.

After five long days, some friends happened to call in to pick up a kid goat from Noel's farm and found him lying on the ground, weak and delirious but thankful to be alive. An ambulance rushed him to Benalla District Memorial Hospital for treatment, but if it hadn't been for his loyal four-legged friends he would have been heading to the mortuary instead.

OUR CHILDHOOD HEROES

The concept of the animal hero has always captured the imaginations and hearts of children, and many of us have grown up with books, television programmes and films that centred on animal protagonists. Here is a trip down memory lane, remembering just some of those all-time favourite animal heroes...

- **BLACK BEAUTY** – the 1877 novel *Black Beauty* by Anna Sewell is one of the best-selling books of all time, with 50 million copies sold. It spoke forthrightly about animal welfare and the importance of treating one another with kindness and respect. But it was perhaps the 1970s TV series *The Adventures of Black Beauty* – a continuation of the story written by Sewell (along with its unforgettable theme tune 'Galloping Home' by Dennis King) – that brought Black Beauty to life for many children, who thrilled at tales of how the brave horse helped to catch criminals and fetch help for people in distress.

- **FLIPPER** – Flipper the dolphin featured first in the eponymously titled 1963 film, then in the subsequent TV series and several other sequel films. Sandy Ricks, who lives in Florida Keys, rescues a dolphin that has been harpooned and the two become inseparable. Flipper's many heroic achievements include rescuing Sandy from a shark and apprehending thugs and criminals.

- **LASSIE** – the much-loved fictional collie dog character was first created by Eric Knight in a short story, and subsequently appeared in a novel, several films and the Emmy-winning television series *Lassie* that aired for 19 years. A recurring theme for the character involved running to fetch help when her master was in trouble or stepping in to fend off a dangerous wild animal.

- **OLD YELLER** – when Walt Disney released *Old Yeller* in 1957, based on Fred Gipson's book of the same name, children and parents alike snuffled into their handkerchiefs at the heart-wrenching tale of poverty-stricken little boy Travis and his best friend, a stray Labrador retriever/mastiff mix, who saves his master from a wild boar and defends the family from a rabid marauding wolf, at the cost of his own life.

- **SKIPPY** – *Skippy the Bush Kangaroo* was a 1960s TV series that achieved international success, bringing the adventures of Sonny Hammond and his friend Skippy, a female Eastern grey kangaroo, into the homes of many delighted children worldwide – except in Sweden, where it was banned because psychologists believed it would mislead children into thinking animals could do things that they cannot. Skippy's exploits on the show included running to get help when Sonny found an injured man, leaping to the rescue when Sonny was held at gunpoint and releasing a grappling hook that got caught around a tree as it was being lowered from a helicopter to lift a car off a man.

Did you know...

... that in the TV series *Lassie*, although the character was female she was usually played by a male dog, because they retain thicker summer coats than female collies, which of course looked better on television?

'Master said, God had given men reason, by which they could find out things for themselves; but He gave animals knowledge... which was much more prompt and perfect in its way, and by which they had often saved the lives of men.'

Anna Sewell

SAVING OTHER ANIMALS

The maternal instinct to protect one's young runs high in most species, and with some this extends to other family or 'pack' members, which in the case of donkeys or horses, for example, might mean another farmyard animal that lives on their territory, while animals such as buffalo might behave in a similar way towards members of their herd. But scientists still remain baffled as to what motivates an animal of one species to intervene in order to rescue an animal of another, whether to save it from a predator or to help in a moment of physical need. When a human does this we call it compassion; when an animal does it we try to tie it back to some instinctive urge, but sometimes it can be hard to make the distinction. Whatever the reasons, the stories of inter-species collaboration in this chapter certainly make for some heart-warming reading.

ZETA'S FRIENDS

A show horse named Zeta received first-class first aid from her field mates after a shocking attack…

In May 2010, Jo Young was at her home in Telford, Shropshire, when she received a call from a man walking his dog near her field to say her horses were acting strangely. Jo, a horsewoman who competed professionally, had several horses out to graze: a gelding, three mares, and Zeta, a 20-year-old bay mare that had competed in show-jumping and dressage events across the UK. The helpful dog walker had noticed the other horses crowded around Zeta, all of them with flecks of blood on their muzzles.

Rushing out to the field, Jo was horrified to see that Zeta was badly wounded – she had been shot twice with a crossbow in what seemed to be a completely random attack. One arrow had bounced off her rib cage to land on the grass, causing only a minor flesh wound, but the other had penetrated and, as Jo later learned, had stopped within an inch of her lung. Jo realised the other horses had instinctively gathered together to nuzzle Zeta for comfort and keep her wounds clean by licking them.

When vet John Brentnall arrived at the scene 30 minutes later he saw the other horses had prevented Zeta's wounds from becoming infected, and had also helped to stem the flow of blood in the three and a half hours since the attack had happened. He was able to remove the bolt and stitch the wounds. But according to him, it was unlikely the horse would have survived without the help of her friends: 'Infections can be lethal to horses but Zeta was extremely lucky to have such good support from the other horses.' Happily, Zeta made a full recovery.

Witnesses had reported seeing a group of teenagers running away from the field at the time of the attack and the police and RSPCA launched an investigation into the cruel shooting. Jo said of the incident: 'There was definitely a herd instinct kicking in among the horses. They knew Zeta was in need and they rallied round to save her.'

CAMEROON GORILLAS

A Cameroonian hunter had clearly underestimated the local gorilla population...

In Yaoundé, Cameroon, the local weekly newspaper *L'Action* published a curious report about a troop of gorillas determined to rescue one of their kin. A group of around sixty of the primates were reported to have entered the village of Olamze, on the border with Equatorial Guinea, in search of a young gorilla that had been captured and brought to the village by a renowned local hunter.

The gorillas turned up just before midnight and walked single file through the village, until shots were fired to frighten them away. The next night they returned, and this time they began to beat angrily on the doors and windows of the houses. Observing the gorillas' determination, the village chief decided to instruct the hunter to release the young gorilla. *L'Action* then stated: 'Immediately, the assailants returned to the forest with shouts of joy, savouring their victory.'

Perhaps the journalist was exaggerating slightly on this final point, but it is certainly charming to think of the gorillas storming a human village in order to take back one of their own.

It's the little things in life

It's extraordinary the lengths some animals will go to, to protect their young. In May 2009, amateur wildlife photographer Dennis Bright captured an astonishing scene at a house in Fareham, Hampshire. A mistle thrush had unwittingly built her nest on a roof close to a downpipe, and when water came gushing down after a heavy rainfall she puffed herself up to twice her normal size to block the water from swamping the nest. The dedicated mother stayed there for hours, getting out every half hour to dry off, and her mate was left with the task of bringing food for her and the chicks, all four of which turned out to be perfectly healthy and flew the nest successfully.

DOTTY

Dotty the donkey proved to be a good friend to have on your side in a tight spot…

Dotty the donkey and Stanley the sheep both belonged to Ann Rogers, 63, and shared a stable on Row Brow Farm, near Scarborough, North Yorkshire. Stanley had been orphaned as a lamb and the pair had become firm friends. One day, when Stanley was in danger, Dotty showed the true strength of her loyalty.

A dog had got into the paddock and attacked Stanley, grabbing onto the sheep and locking its jaws. As soon as Dotty saw what was happening, she raced down the field to help her distressed friend. She pinned the dog to the ground until it let go of the sheep, which lost two teeth and suffered facial paralysis due to the attack. Once the sheep had recovered from the ordeal, she was never far from Dotty's side.

Dotty was later presented with an award by the PDSA for her bravery. Elaine Pendlebury, a PDSA vet, said that she found the donkey's behaviour outstanding. Donkeys often react defensively when faced with a threat, and in this case 'Dotty showed herself to be a true protector of the animals she sees as her family'.

PADDY

In the midst of a terrifying bushfire, policeman Mike
Salmon was glad to have an old colleague at his side…

The bushfires which came to be known as Black Saturday
raged across the Australian state of Victoria in February
2009, resulting in a huge loss of life and widespread
destruction. On the eve of Black Saturday, Mike Salmon
noticed that the grass around his house in Happy Valley
near Myrtleford was dry, 'like talcum powder'. By 6.30
p.m. on the Sunday, Mike could see the bushfire spreading
up through the valley towards his home. He began to do
what he could to avert disaster, patrolling his property for
drifting embers and hosing his house down with water.
But before long the fire had jumped across the road and
taken hold on the other side of his house. When the two
banks of fire joined behind his house, Mike realised he was
surrounded.

Seeing that the fences around his property were alight,
Mike's worst fear was that his animals would be burned
alive. He had four sheep, two goats and a beloved retired
police horse named Paddy that had served alongside him
on parades at police graduations and other major events.
The two had formed a very special bond, having worked
so closely together over the years. Mike decided to let the
sheep and goats out of their pens; he knew there was a risk

they might flee into the path of the fire, but the thought of leaving them trapped and helpless as the flames burned ever closer was much worse.

Mike continued to patrol his house, and when he next came back to the area where he had released the animals he was surprised to find Paddy had somehow rounded them all up, and that they were cowering for safety under his huge frame – and the horse seemed totally calm. Mike instructed Paddy to stay put and look after the other animals, then continued to patrol his house as the fire raged on, coming back to check on them at half-hour intervals and to pour water over Paddy's flanks to keep him cool. Despite the roaring flames nearby and the embers falling from the sky, Paddy never once showed signs of panic. If the other animals moved away, he would round them up again.

Later, Paddy was nominated for a bravery award by the RSPCA, for his actions during the fires, which he miraculously survived with only a small burn on his nose. Mike would never forget his horse's bravery, although he also said he couldn't help being a little bemused that Paddy would strive to protect the other farm animals: 'He's a bit superior to them; he doesn't like them. They're just sheep and goats and he doesn't have much to do with them.'

BEAUTY

In the days before men began to train dogs for search and rescue purposes, a brave terrier showed a natural aptitude for finding other animals in distress…

During World War Two, when London was being bombed regularly by the enemy, the PDSA sent out animal-rescue squads to search for any animals that may have been trapped under rubble. Superintendent Bill Barnet was a member of one such squad who always used to take his dog Beauty along with him for company when he was out at work.

One night in 1940, the little wire-haired terrier suddenly decided to join in of her own accord, and began digging furiously in one spot. Some minutes later the rescue squad helped Beauty to uncover what she was looking for – a terrified, but otherwise unharmed, cat that was buried in the rubble beneath a table.

From then on, Beauty continued to help Bill Barnet on rescue missions, and by the end of her career she was credited with having saved the lives of 63 animals. She had no training, and is generally recognised as being the first official search and rescue dog. Towards the end of the war, the authorities began training dogs specifically to trace buried casualties. Because she scrabbled so hard with her little paws when uncovering a find they were often sore and bleeding, and admirers sent her some specially made leather boots to help protect her feet.

In recognition of her work, Beauty was awarded the PDSA Pioneer Medal, which is normally only given to people. She also received a silver mounted collar inscribed 'For Services Rendered' from the Deputy Mayor of Hendon, and was granted 'the freedom of Holland Park and all the trees therein!' In January 1945 she was honoured with the PDSA Dickin Medal: 'For being the pioneer dog in locating buried air-raid victims while serving with a PDSA rescue squad.' Search and rescue dogs are now used all over the world in various situations, including saving human lives.

ZULU ELEPHANTS

A herd of elephants proved that community spirit is well and truly alive in their game park in South Africa…

In 2003, a private game company was at work on the Thula Thula Exclusive Private Game Reserve in Empangeni, Zululand, where they were capturing antelope to be relocated for a breeding programme. The antelope were rounded up into a boma, a type of temporary enclosure, where they would be held until the time came to move them.

As the team settled down for the night, they were disturbed by some uninvited visitors arriving at the camp. A herd of 11 elephants sauntered in and made their way purposefully towards the boma. At first the crew thought they had been attracted by the smell of the lucerne (alfalfa) that they had been keeping to feed to the antelope. But it soon became apparent they had other ideas.

As the herd circled the boma the staff watched them warily. Then the herd matriarch, known as Nana, approached the gate of the boma and, using her trunk, she calmly and carefully undid all of the latches holding the gate closed. As the gate swung back she stepped to the side – it was at this point that the onlookers realised the herd were not on the hunt for a midnight snack, but on a rescue mission. The elephants stood by as the antelope escaped from the boma and disappeared off into the night, then left just as silently as they had come.

Commenting on the incident, an ecologist named Brendon Whittington-Jones said: 'Elephants are naturally inquisitive – but this behaviour is certainly most unusual and cannot be explained in scientific terms.'

Saved at sea

On three separate occasions scientists onboard the *Golden Fleece*, in the seas near the Antarctic Peninsula, observed humpback whales protecting a seal that was being hunted by killer whales, with one instance of a humpback rolling a seal out of the water onto its chest and arching its back to lift it out of reach of the orcas. The scientists speculated that the menacing behaviour of the killer whales had triggered a protective maternal response in the humpbacks, causing them to act instinctively to counter the threat posed to a smaller animal – but we can never know for sure.

MOKO

A dolphin had an important role to play in a drama
that unfolded off the coast of New Zealand…

In March 2008 in New Zealand, Department of Conservation worker Malcolm Smith was alerted by a local man that two pygmy sperm whales, a mother and her calf, had been stranded on Mahia Beach, about 300 miles north-east of Wellington on the east coast of the North Island. Around thirty whales are stranded on this stretch of coast annually and sadly most have to be killed. Knowing this, Malcolm rushed out with several volunteers to attempt to rescue the whales.

Down on the beach, the team kept the whales wet and worked for over an hour to refloat them so they could direct them back into open water, but they kept on getting stuck on a large sandbar just offshore. The whales were becoming increasingly disorientated and tired, and as he listened to their pitiful distress calls Malcolm began to think he might have to take the sad decision to put them both down, and save them from a long and painful death.

But just then someone else answered the whales' calls. Splashing through the water came a dolphin, known by the locals as Moko. As soon as she arrived the whales re-submerged in the water, and Moko swam between the rescuers and the whales and began to lead them 200 yards along the beach, then out through a channel to the open sea.

In 30 years in his job, Malcolm had never seen anything like it. 'The things that happen in nature never cease to amaze me,' he marvelled. 'I was not aware dolphins could communicate with pygmy sperm whales, but something happened that allowed Moko to guide those two whales to safety.'

After her emergency rescue work was done, Moko returned to the beach to play in the surf with the locals. Scientists believed she had become separated from her pod, and so had settled in the area, where she was well known and was often seen playing with swimmers, approaching boats to be patted, and pushing kayaks through the water with her snout.

Anton van Helden, a mammals expert at New Zealand's national museum, Te Papa, said it was the first time he had ever heard of 'an inter-species refloating technique'.

A helping flipper

Dolphins also saved the day in September 1983 at Tokerau Beach, New Zealand, when a pod of 80 pilot whales were beached and left stranded by the ebbing tide. Local people waded out, talked soothingly to the whales and kept their skins wet. When the tide came back in they managed to refloat the whales and had turned them around in order to direct them to deeper waters when a school of dolphins swam into the shallows, surrounded the whales and guided them out to sea, saving 76 of the pilot whales' lives.

ANIMALS IN THE LIMELIGHT

The online video-sharing community YouTube is a gold mine of animal clips – but it's not just about talking dogs, sneezing pandas and dramatic gofers. Some of the most extraordinary footage of animals helping each other or displaying uncharacteristic behaviour towards other species has been captured – be it by professional wildlife journalists or on dad's camcorder – and posted on the site. Visit www.youtube.com and type the key words highlighted in bold below into the search box to watch some truly impressive animal magic moments...

- **HERO DOG SAVES DOG HIT BY CAR** – a dog braves oncoming traffic to drag another dog that has been hit by a car off the highway. Fire fighters spot them and help, and the injured dog's life is saved.

- **DOG STOPS BULL FROM KILLING MAN** – a man in a bullfighting ring has lost the upper hand and is on the ground being savaged by a bull. Out of nowhere comes a stray dog that terrorises the bull, distracting it so that the man can escape.

- **HIPPO SAVES ANOTHER ANIMAL** – a crocodile snatches a young impala drinking at the water's edge and is dragging it under when a hippo charges over, frightens the croc off and drags the impala up onto the bank. It gently takes its head in its jaws, as if trying to revive it. Sadly the impala later dies and the croc returns to claim his meal.

- **LEOPARD CUDDLES BABY BABOON** – stunning National Geographic footage of a leopard that kills an adult female baboon, then realises that a live baby monkey is still attached to the dead mother's back. The leopard protects the baby from an approaching hyena by taking it up a tree, then plays with it gently and licks it.

- **LION ADOPTS A BABY ANTELOPE** – a lioness cares for a baby antelope, normally a perfect snack for a hungry lion.

- **BATTLE AT KRUGER** – this incredible video has received over 60 million hits, and when you watch it you'll know why. Lions take on a herd of buffalo at a watering hole, and a calf is separated and slips into the water. When the lions try to drag the calf out of the water a crocodile grabs it. Eventually the lions win the tug of war and are preparing to kill and eat the calf when the herd attacks the pride and the calf escapes.

WATER RESCUE

Although humans have crossed great oceans on ships, ridden massive waves on surfboards and swum the length of the world's longest rivers, the water is not our natural domain, and the stories in this chapter give some measure of just how vulnerable we can be when you consider the many dangers of the watery world. Not surprisingly, this is where the dolphin has its day, with several stories about them coming to the rescue, whether defending humans from dangerous underwater predators, saving us from drowning or guiding us to safety. Other marine mammals are reported to have pitched in and lent a helping flipper, too. Dogs are known for their strong swimming ability and have been credited with many watery rescues – the dog that appears in this chapter set something of a record, and you will meet one rather more unexpected lifeguard that hails from the farmyard.

BOTTLENOSE DOLPHIN POD

When Todd Endris set off one morning for a day's surfing off the coast of California he had no idea of the danger that was waiting for him beneath the waves...

On a late August morning in 2007, surfer Todd Endris, aged 24 and the owner of Monterey Aquarium Services, headed into the waters of Marina State Park, off Monterey, California, with friends to catch some waves. Just before 11 a.m., Todd was sitting upright on his surfboard when a great white shark 'came out of nowhere'. He estimated the shark was 12 to 15 feet in length which, given its terrifying surprise attack, just goes to show what efficient predators these huge animals are. Thankfully, it didn't manage to get its enormous mouth around both him and the board.

Coming in for a second time, however, the shark clamped its jaws onto his torso, sandwiching the board and Todd in its mouth and literally peeling the skin off his back 'like a banana peel', as Todd later described it. Luckily, because his stomach was pressed against the surfboard, his intestines and internal organs were not damaged. Round three – and this time the shark latched on to Todd's right leg. As the shark's grip held him in place he kicked at its head and snout with his left leg until it let go. Now he was seriously weakened, bleeding profusely and far from the shore. If it wasn't for what happened next, the shark would have certainly killed him.

A pod of bottlenose dolphins that had been playing in the surf a short distance away swam over to Todd and began to circle him, keeping the shark at bay. Within the safety of this protective ring he was able to summon the strength to clamber back on his board and ride a wave back onto the shore to safety.

There, he was administered first aid until a Medivac helicopter arrived to fly him to hospital. Blood pumped out of his right leg, which had been bitten to the bone, and it was probably the actions of the quick-thinking first aider that stopped him dying from further blood loss: the man, named Simpson, used his surf leash as a tourniquet to stem the blood flow. At the hospital, the surgeons had a big job on their hands, figuring out how to put Todd back together.

Just six weeks later, Todd was well enough to go surfing again – and he went back into the water at Marina State Park. It wasn't easy, but he had to face his fears. As for the shark, it was protected within the marine wildlife refuge of the park, although Todd said he wouldn't want the shark to be harmed in any way, 'We're in his realm, not the other way around,' he said. Nearly four months later he was still undergoing physical therapy to repair his muscle tissues, and in an interview with *Today*'s Natalie Morales he said that the dolphins' rescue was 'truly a miracle'.

PELORUS JACK

A Risso's dolphin dubbed Pelorus Jack became famous for escorting ships safely through the dangerous waters of French Pass in the Cook Strait, New Zealand...

The Risso's dolphin that came to be known as Pelorus Jack was first seen in 1888 when he appeared in front of the schooner *Brindle* in the waters near French Pass. This treacherous channel, used by boats travelling between Wellington and Nelson, is situated between D'Urville Island and the South Island, and has claimed many shipwrecks over the years due to the rocks and strong currents. Between 1888 and 1912, Pelorus Jack continued to guide boats safely through the area – no boat that he accompanied was ever shipwrecked. He would guide ships by swimming alongside them for 20 minutes at a time and, if a crew could not see Jack before they entered the channel, they would often wait for him to appear.

The dolphin was usually spotted in Admiralty Bay, between Cape Francis and Collinet Point, near French Pass, and despite his name was not a resident of Pelorus Sound. He was around 13 feet long, white in colour with grey lines, and had a round, white head. The Risso's dolphin is uncommon in New Zealand's waters, and only 12 Risso's dolphins have ever been recorded in the area Jack used to frequent. Pelorus Jack was sighted by many sailors and travellers over the years, and he was mentioned in local newspapers and depicted in postcards.

In 1904, a passenger aboard the SS *Penguin* shot at Pelorus Jack with a rifle. But, in spite of this attempt on his life, Pelorus Jack continued to guide ships, although the story goes that he would no longer guide the *Penguin* after this incident – and it was shipwrecked in 1909. Following the shooting, Pelorus Jack became protected by Order in Council under the Sea Fisheries Act on 26 September 1904, and it is believed he was the first individual sea creature to be protected by law in any country.

The famous dolphin was last sighted in April 1912. Some feared foreign whalers had harpooned him, but (as his pale body colouring indicated) he was an old animal and it is more likely he died of natural causes. In any case, his story has stood the test of time, and since 1989 his image has been used as a logo by the Interislander ferry service that regularly crosses the Cook Strait.

POD OF DOLPHINS

A group of lifeguards on a training swim off the coast of New Zealand had an unusual interruption to their lesson…

On 30 October 2004, lifeguard Rob Howes accompanied three female lifeguards on a training swim about 300 feet off Ocean Beach, near Whangarei on the North Island of New Zealand.

As they were swimming the group noticed a pod of dolphins approaching quite quickly. The pod began to circle the swimmers closely, keeping a distance of only 2–4 inches and bunching the four of them together. They kept doing this for around 40 minutes, slapping the water with their tails vigorously.

Howes felt a little unnerved by the speed of their approach and their behaviour, and wondered if it was a group of aggressive males, or perhaps female dolphins trying to protect their young.

At one point there was an opening in the ring, and Howes drifted away from the others – which was when he saw a great white shark only 6 feet away. Howes remembered there had been a number of sightings in the area around that time of year, as they came into the harbour to give birth.

As the shark started to swim towards the other swimmers, he watched the dolphins become even more frantic. He later said: 'I would suggest they were creating a confusion screen around the girls. It was just a mass of fins, backs and human heads.' A rescue boat eventually arrived, at which point the shark swam away, but the pod of dolphins stuck close to the group as they swam back to shore.

MILA

In 2009, a girl who was free-diving in an arctic pool found her leg clamped in the beak of a beluga whale…

Athlete Yang Yun was taking part in a competition at Polarland, a huge polar-themed aquarium in Harbin, a city in northeast China. Contestants had to free-dive to the bottom of a 20-foot Arctic pool that was home to the attraction's beluga whales – and stay down as long as possible. But due to the freezing temperature of the water in the pool, the 26-year-old's legs began to cramp, leaving her immobilised in the water and unable to surface for air. At such depths, water pressure will keep a body underwater, especially if the person is unable to move and push themselves upwards.

Luckily for the girl, a beluga whale named Mila sensed her distress and, taking her leg in its beak, pushed her up towards the surface. As belugas feed only on small fish and squid they do not have large teeth, so the girl was unharmed. Speaking of the frightening experience, the diver said: 'I began to choke and sank even lower, and I thought that was it for me – I was dead. Until I felt this incredible force under me driving me to the surface.'

An official at Polarland said Mila had spotted the girl was in trouble before anyone else did, and described her as a sensitive animal who works closely with humans. Belugas are intelligent creatures, and have facial muscles that allow them to form expressions that look like very real smiles. Almost certainly, Yang Yun owed the beluga her life.

PRISCILLA

You might be more familiar with the image of a pig wallowing in the mud than splashing through open water, but one pig proved to be a particularly strong swimmer…

Victoria Herberta was the proud owner of Priscilla, a pig that used to follow her around on a lead like a pet dog. One day in 1984, Priscilla's owner took her out for the day to meet friends at a lake in the Houston area of Texas. The three-month-old pig, wearing her purple harness and lead as always, decided to have a dip in the water.

Among other visitors to the lake that day was 11-year-old Anthony Melton. He was wading out into the lake when he lost his footing on a drop-off ledge and plunged into deeper water. Unable to swim, he began to struggle and panic. Seeing Priscilla swimming nearby, Anthony's mother shouted to her son to reach out and grab the pig's lead. The boy did – and then the pair disappeared beneath the water. For a few terrifying moments there was no sign of them, until they resurfaced and the 45-pound pig struck out for the shore, towing the boy, who was twice her own body weight, for more than 150 feet.

After her heroic feat that day, Priscilla became the first member of the Texas Animal Hall of Fame and was declared a local heroine. She made appearances on television, and was chauffeured to the University of Illinois to open the International Belly Flop Contest. She was also honoured with the American Humane Society's prestigious Stillman Award, named after Dr William O. Stillman, who was known for his efforts to protect animals from cruelty.

SWANSEA JACK

The black Labrador retriever dubbed 'Swansea Jack' made as many as 27 watery rescues during his short life in the 1930s...

Jack lived with his owner William Thomas at Padley's Yard, Wales, on the western bank of the River Tawe – an area that was made derelict after Swansea's shipping industry shifted to the eastern side of the river.

In 1931, aged just one, the Labrador retriever made his first heroic rescue when a 12-year-old boy, who was playing on the wharf, fell into the water. As a puppy, Jack had always been frightened of deep water (perhaps explaining why he was so watchful of humans in the water), but as soon as he saw the boy was in trouble he jumped in and dragged him back to the shallows, where the boy struggled ashore. Despite his timely action, Jack's bravery was not reported at the time.

Several weeks later, Jack performed a second successful rescue by saving a flagging swimmer from the nearby waters of North Dock. This time his actions attracted a small crowd, and his photo and an account of the rescue were printed in the local newspaper. He was awarded a silver collar by the city council for his efforts, and sprang to fame as a local hero.

By the age of five, Jack had made so many rescues that he was featured in the national newspapers. He won numerous medals for his service to humans, including two bronze stars from the National Canine Defence League, the Bravest Dog of the Year award from both the *Daily Star* newspaper and the *Daily Mirror* in 1936, and the Bravest Dog category at Crufts. He was taken on a nationwide tour and was presented with a silver cup by the Lord Mayor of London. Later, Jack also helped to raise substantial amounts of money for charitable causes, when his owner permitted the famous and wealthy to be photographed with him.

Sadly, on 2 October 1937, aged just seven, Jack died after accidentally eating rat poison. A memorial to this charismatic and courageous canine was erected near his favourite swimming spot on the promenade in Swansea, near St Helens Rugby Ground. It can still be visited today.

SEAL

Humans aren't the only type of mammal to have needed rescuing in the water. Although most dogs are strong swimmers, one injured canine found itself in trouble in strong currents in the River Tees…

Chris Hinds was out walking his dogs with his 18-year-old son, Raymond, near Newport Bridge in Middlesbrough, at around six o'clock one evening, when he saw an injured dog on the bank of the River Tees. The Labrador-German shepherd cross had a cut on its head, but when Hinds reached towards the animal it flinched and plunged into the river, where it was swept away by the strong current.

The dog seemed to be really struggling as he followed its path down the river. It was 30 feet out and there was no way Hinds could reach it – he felt sure the poor creature was about to drown. But just as the dog began to disappear beneath the surface a seal appeared and began to circle the animal, and nudge it back towards the mudflats. 'I have never seen anything like it in my life,' said Hinds.

In the meantime, Raymond had run to their car parked nearby and used his father's mobile phone to call the fire brigade. Sub-officer Mark Baxter of Stockton Fire Station arrived with his team in time to see the dog back on dry land and three seal heads bobbing out of the water, just off the bank. The rescuer seal and his two friends stayed to watch as the fire crew captured the confused and sodden dog, which was handed over to the RSPCA until its owner could be found. Hinds later commented: 'This dog would not have survived in the water if it hadn't been for that seal.'

Dominic Waddell, an expert in seal behaviour and senior aquarist at Scarborough's Sea Life & Marine Sanctuary, explained that common seals are instinctively protective, highly inquisitive animals, and rarely aggressive. He suggested the seal might have seen the dog as something unusual which should not have been in the water, then pushed it towards the river bank because it sensed it should be on land.

OUR AQUATIC COUSINS?

'The happiness of the bee and the dolphin
is to exist. For man it is to know
that and to wonder at it.'

Jacques-Yves Cousteau

Dolphins have played a role in human culture ever since ancient Greek times, when they appeared on coins pictured with a boy or deity riding on their backs, and when a dolphin riding the wake of a ship was considered a good omen. In the stories, tales of dolphins rescuing drowning sailors and guiding ships through treacherous waters abound. For Hindus, the Ganges River Dolphin is associated with Ganga, the deity of the Ganges. In modern popular culture they are typically represented as heroic, intelligent creatures, such as Flipper of the 1960s television series, and in Douglas Adams' *The Hitchhiker's Guide to the Galaxy*, in which dolphins are ranked as the second most intelligent species on Earth (after mice – and before humans!).

Perhaps one of the reasons for dolphins' enduring popularity is the many perceived shared characteristics between the cetaceans and humans...

- **AS WELL** as being able to communicate efficiently with each other through whistles and clicking sounds, research carried out by Dr Vincent Janik of the Sea Mammal Research Unit at St Andrews University found that dolphins also have names – an individually distinctive signature whistle is developed in the first few months of life and appears to be used in individual recognition.

- **DOLPHINS** love to play, surfing swells, leaping about in the wakes created by boats, performing acrobatic jumps out of the water and interacting with swimmers.

- **THE MANY REPORTS** of dolphins rescuing humans at sea, and the mystery behind what their possible motives could be, have led experts to speculate whether they are the only animals besides humans to exhibit altruism.

- **DOLPHINS** are known to have sexual intercourse for reasons other than reproduction, and homosexual behaviour has also been observed in the species.

Did you know...

... that the average bottlenose dolphin's brain weighs 1,500–1,600 grams, compared with the average human's at 1,200–1,300 grams?

'When I see a dolphin,
I know it's just as smart as I am.'

Don Van Vliet

INCREDIBLE SURVIVORS

The animals in this chapter are a true inspiration – through their courage, triumph over suffering and adversity, and sheer will to survive they are in many ways symbols of hope to us all. It is stunning what some hardy animals can endure – take, for example, the horse that received horrific injuries in a terrorist attack, or the tiny rodent that went through a recycling plant. Others – such as the dog that survived a tsunami and the cat that survived the London Blitz – are a source of hope because of their survival in conditions where many other animals, and humans, have fallen.

TSUNAMI DOG

A dog became the focus of an inspirational rescue
in the aftermath of the 2011 earthquake in Japan…

When an earthquake of 9.0 magnitude struck off the coast of Japan in March 2011 – the most powerful ever recorded in the country – there was widespread destruction caused by the massive tsunamis that followed. Whole villages were decimated, thousands of people disappeared and reactors at the Fukushima Daiichi Nuclear Power Plant went into meltdown.

The rescue operation, in which search teams looked for people in the rubble, was a long and gruelling affair, with only a few success stories. Three weeks after the disaster, a coast guard helicopter was out looking for missing people and spotted the roof of a house about a mile off the coast of Kesennuma, Miyagi. Those onboard didn't hold out much hope but went to investigate in case anyone had been trapped inside the submerged house. Going in closer they were amazed to see a medium-sized brown dog trotting around the roof, before disappearing inside the house through a broken section.

One of the team was lowered down to investigate – perhaps the dog's owners were still alive inside? The rescuer tried to entice the dog out, but when that failed he entered the house himself by ripping a wider opening.

There was no one inside but thankfully managed to rescue the dog, who had survived alone at sea for three weeks. In such tragic times, even the smallest life saved can seem like a miracle and produce a spark of hope.

COMANCHE

Comanche's extraordinary tale of survival made him one of the most famous battle horses in history…

Comanche was a mustang/Morgan bay gelding acquired by the US army in 1868 in St Louis, Missouri. He stood at 15 hands and had a small white star on his forehead. Along with other horses, he was sent to Fort Leavenworth, Kansas, before moving to the 7th Cavalry's camp near Ellis. He became the preferred battle mount of Captain Myles Keogh, and was injured several times – including one occasion when he was shot with an arrow through his right hind-quarter, yet still bravely carried on, while fighting the Comanche Indians, and it is believed this is where his name came from.

On 25 June 1876, Comanche was ridden into the Battle of the Little Bighorn. This infamous battle, also known as Custer's Last Stand (and as the Battle of the Greasy Grass by the Indians) took place in eastern Montana Territory when combined forces of the Lakota, Northern Cheyenne and Arapaho peoples came to blows with the 7th Cavalry Regiment, a force of 700 men led by George Armstrong Custer. It was the most famous action of the Great Sioux War and an overwhelming victory for the Indian forces. Five of the 7th Cavalry's companies were obliterated, including General Custer himself.

Of the horses that went into battle and survived that day, it is believed many were captured by the Indians but Comanche was too seriously injured to be of any use to them – close to death, he lay in a ravine, his poor body shot through in several places with bullets. Two days later he was discovered by Sergeant Milton J. DeLacey – he was too weak to even walk and was taken to Fort Lincoln, where he was given great care and attention and made a slow recovery.

On 10 April 1878, Colonel Samuel D. Sturgis issued an order declaring Comanche should never again be ridden or do work of any kind. He said of the horse: 'Wounded and scarred as he is, his very existence speaks in terms more eloquent than words, of the desperate struggle against overwhelming numbers of the hopeless conflict...' Comanche was also honoured with the title of Second Commanding Officer of the 7th Cavalry.

In June 1879 the by-now famous horse was taken to Fort Meade, where he lived a life of luxury until 1887, and was then moved to Fort Riley, Kansas, where he became a favoured pet, and was indulged in his fondness for beer and sugar lumps. He appeared in parades, including the official regimental mourning ceremony each 25 June, when he would be draped in black with stirrups reversed. Comanche died peacefully in November 1890 and is one of only two horses in US history to have been given a funeral with full military honours.

Buried with honours

Black Jack was the only other horse to have been buried with full military honours in US history. A coal-black Morgan-American Quarter Horse cross, he served in the Caisson Platoon of the 3rd US Infantry Regiment (the Old Guard) and was named after General John J. ('Black Jack') Pershing. He was the riderless horse in more than 1,000 armed forces full honours funerals, at which he appeared with boots reversed in his stirrups to represent the symbol of a fallen leader. Among the highlights of his long and respectable military career were the state funerals of presidents John F. Kennedy, Herbert Hoover and Lyndon B. Johnson, and General Douglas MacArthur. He died on 6 February 1976 and his cremated remains were laid to rest in a plot at Fort Myer, Virginia, on Summerall Field, 200 feet northeast of the flagpole in the southeast corner of the parade field.

FAITH

When St Augustine's Church in London was hit in an air raid during World War Two, a brave little tabby named Faith stayed to protect her kitten, even as the building collapsed around her…

Faith first arrived at St Augustine's Church as a stray, looking for food and warmth. She was turned out three times by the verger, but when the rector, Henry Ross, saw her he took pity. When nobody came to claim the cat Ross decided to keep her and named her Faith. She became a popular feature of church services, lying stretched out at Ross's feet or on the front pew during the sermons.

In August 1940, the parish welcomed another new arrival into the fold when Faith gave birth to a black-and-white tom kitten, named Panda. The two were settled into a basket in the rector's living quarters and getting on very well when Faith began to behave strangely, investigating the different rooms of the house and appearing very restless. One day she took the kitten by the scruff of its neck and moved it down to the basement. When Ross found them there he moved the kitten back upstairs where it was warm. But the next day Faith had moved Panda back to the basement. This happened three times, until Ross gave in and decided to respect Faith's wishes by moving the basket down to the basement, where the pair happily settled in-between stacks of music sheets.

On 9 September, Ross had to go to Westminster on business. As he made his way home that evening, the air-raid warning was given and he had to spend the night in a shelter. That night the bombing was very severe and many buildings were destroyed, including eight churches. When he returned home, only the tower of St Augustine's was still standing – his home had been reduced to a mass of rubble, some of which was still on fire. Firemen warned him to move away from the scene, telling him no one could have survived the blast, but the resolute Ross approached the debris anyway, still holding out hope for his beloved cats.

A faint meowing sound came from beneath a pile of rubble and timbers. Ross struggled to move the debris aside, revealing two dirty, bedraggled, frightened but completely unharmed cats. Relieved and thankful, Ross took the two cats to the safety of the church vestry, which remained intact.

Faith's astonishing story was reported in the papers and, as the news spread, many tributes came in. As a civilian cat Faith was not eligible to receive the PDSA Dickin Medal – an honour reserved for military animals – but founder of the award, Maria Dickin, did present her with her own special silver medal in recognition of her steadfast courage. She was the first cat to receive such an accolade for bravery.

Perhaps some would argue that what Faith did is not so extraordinary – after all, it is a natural instinct for a cat to protect her young. But at a time when Britain was under attack and hundreds of people had died in air raids, or lost their homes, the story of this little creature and her kitten surviving against all odds would have brought hope, not just to the local community that had seen its church destroyed, but to the many others who heard her story as it spread through the land. Henry Ross had got it spot on when he named her Faith – she had brought the people of Britain just that.

SEFTON

The survivor of a horrific terrorist attack became a national symbol of courage and resilience...

On 20 July 1982, the new guard of the Blues & Royals Mounted Squadron left the Hyde Park barracks and rode out towards Whitehall, where they were due to perform the Changing of the Guard. The beautiful, glossy black horses, with their uniformed mounts, made quite a picture as they stepped boldly out in the morning sunlight. But on this day the ritual changeover was not to go smoothly. As the convoy approached Hyde Park Corner, admired and applauded by onlookers who lined the route, a nail bomb planted in a nearby car by the IRA detonated. The bomb contained up to 10 pounds of

explosives, girdled by hundreds of 4- and 6-inch nails, designed to inflict maximum injury. Four members of the guard and seven horses were killed in the blast, and three horses were left severely wounded.

One of the horses that survived the explosion was Sefton, a 19-year-old black horse that stood out because of his white socks and blaze. Sefton's jugular vein was severed, a 6-inch nail pierced his head, one of his eyes was burned and damaged, and his body was studded with 28 pieces of shrapnel and car body. He suffered severe shock and loss of blood, and yet his rider Trooper Pederson later said, in spite of the injuries, the brave horse didn't try to throw him in the moments after the bomb went off.

In total, Sefton had 38 wounds to his body, and it was to be a long road to recovery, starting with a gruelling eight hours of surgery. The veterinarians treating him gave him a 50–50 chance of survival, and yet he somehow managed to pull through. Get well soon cards, mints and donations flooded into the hospital from well-wishers inspired by this brave horse – Sefton had become a symbol of courage and resilience for members of the public who had been shocked and outraged at this vile attack targeting innocent, noble creatures. The sum of donations was eventually enough to construct a new surgical wing at the Royal Veterinary College, which was named the Sefton Surgical Wing.

Incredibly, Sefton not only made a full recovery and returned to work with his regiment, but was able to pass the very spot where the tragedy had occurred with his head held high and not the slightest sign of panic. By contrast, some of the other horses injured in the incident had their nerves affected permanently, and would jump at the sound of a stable door slamming. He was awarded Horse of the Year, and he and Pederson received a standing ovation at the show.

On 29 August 1984, Sefton was retired and went to live out the rest of his days in peace at the Home of Rest for Horses, where he was the favourite of the staff and visiting public. He died, aged 30, on 9 July 1993, and his marble headstone overlooks the fields at the Defence Animal Centre, in Remount Road, Melton Mowbray.

ALDANITI

One very special horse-and-rider team captured the hearts of the nation at the Grand National in 1981…

Aldaniti was a racehorse named after his breeder Tommy Barron's four grandchildren: Alastair, David, Nicola and Timothy. He was born in the summer of 1970 in Yorkshire and was later purchased by Nick Embricos, who lived in Sussex. He was trained by Josh Gifford and showed some promise, especially at jumping, but sadly he was constantly troubled by problems with his legs. After one particular injury he was stabled for seven months, and with a serious tendon injury in 1979 it looked like he was unlikely to race again – and he certainly didn't seem like a contender for the Grand National. Yet, despite all these problems, he was a calm and gentle horse, and his trainers found him a pleasure to work with.

Aldaniti wasn't the only one whose career was hanging in the balance. Jockey Bob Champion had enjoyed a successful career so far and he was set on winning the Grand National – but in 1979 he was diagnosed with testicular cancer. Doctors predicted that he had only a 40 per cent chance of recovery, and that if he didn't recover he would have a maximum of eight months to live. An aggressive course of chemotherapy was begun immediately but, rather than take time off work during this exhausting period, Bob kept on training and racing, with the 1980 Grand National always in his sights. The treatment took its toll, however, and a large scale infection took Bob out of the game for a while, forcing him to shelve his Grand National dreams.

Once he was back on his feet, Bob began training again in earnest, and now had a fellow survivor to work towards his goal with in the form of Aldaniti. Bob realised straightaway that the horse had true spirit, after riding him for the first time in Leicester, saying: 'This horse will win the Grand National one day.' It was as if the two gave each other the will to keep going – Aldaniti struggling with leg injuries and Bob with his cancer treatment in the run up to the 1981 Grand National. Just a week before the big race, Aldaniti was almost hit by a speeding car, and three days before the horse was due to be transported to Aintree a foot-and-mouth scare looked set to close off the area.

In spite of all the obstacles that had faced them, the pair were there at the starting line on race day in April 1981. And what a race it was – there can't have been a dry eye in the stadium when outsiders Bob Champion and Aldaniti raced to victory, coming in four lengths ahead of favourite Spartan Missile. Back home in Findon, 3,000 well-wishers awaited the victorious duo, and the stable was swamped with cards, flags and telegrams. Having fought against seemingly insurmountable odds they had not only won the race – they had won a victory over pain and suffering, and become symbols of hope and inspiration to others.

Aldaniti was subsequently to become a regular at fundraising events, and even played himself in the film *Champions*, starring John Hurt. The jockey went on to run the Bob Champion Cancer Trust, raising millions of pounds for cancer research. Aldaniti died peacefully of old age in 1997; as legends of the horse-racing world, their legacy lives on today.

MIKE

Thanks to advancements in technology many common household items can now be recycled, but no one ever intended this to include family pets…

At Recyclo recycling plant in Flintshire, North Wales, staff were surprised to find a hamster coming through the system along with other waste items. In a four-minute journey that must have resembled a scene from *Toy Story 3*, it passed through a giant shredder used to break down items as large as washing machines, a rotating drum and vibrating grids, before being discovered in a sorting area. Amazingly, the tenacious little creature survived practically unscathed, with only a sore foot to show for his ordeal.

It is thought he had fallen into a skip of rubbish that was brought into the plant, and from there was transferred into the recycling system. The plant deals with around 300 to 400 tonnes of dry waste daily from all over Cheshire, Flintshire and Wrexham. Some of the material is shredded, before passing along a series of conveyor belts and grids that enable smaller pieces of waste to fall through. Plant manager Tony Williams guessed the hamster was small enough to pass through the blades of the shredder unharmed, but too big to fall through the holes in the trammel. This terrifying journey would have been enough to finish off many animals from sheer shock, but the little hamster tenaciously clung onto life.

Plant worker Craig Bull took the hamster with an inspiring story to tell home to his son Liam, who named him Mike, and gave him a happy new home where he would be kept away from recycling bins. Liam said: 'I can't believe he's still alive after what happened, but he's doing fine now.'

ANIMAL INSPIRATION

Since early men began daubing primitive images of animals onto the walls of their caves and told stories of the animal spirits around the fire, animals have been the source of inspiration for and the subject of the arts throughout the centuries. In ancient times, Egyptians depicted many of their gods with animal heads and had sculptures of cats, which they worshipped; while in the Middle Ages animals and mythical beasts adorned the pages of manuscripts and tapestries. In the seventeenth century poet La Fontaine's animal stories became widely popular and by the eighteenth century the Brothers Grimm's fairytales famously included many creatures. From then on to the present day animals have featured in books such as *The Wind in the Willows* and *Alice in Wonderland*, in music such as 'Peter and the Wolf' and films, including *The Lion King*. Here are three animal heroes that have been made immortal through the arts of writing, painting and music...

- **EACH** of the animal stories in the well-known *Aesop's Fables* illustrates a moral lesson. They are credited to Aesop, a slave and storyteller said to have lived in ancient Greece between 620 and 560 BC, and are still a popular choice for children today. In one of the fables, 'The Tortoise and the Hare', the tortoise triumphs over the hare in a race by progressing slowly and steadily towards the finish line, while the hare, so sure of his own success because of his superior speed, takes a nap and is overtaken by the determined tortoise.

- **ENGLISH PAINTER** George Stubbs (1724–1806) was best known for his paintings of horses. He had studied anatomy and his knowledge of equine physiology was unsurpassed. One of his most famous paintings was *Whistlejacket*, in which a chestnut stallion is depicted rearing up in almost photographic detail against a plain background. The subject was a real racehorse, known for his fiery spirit, that was beaten only four times in his racing career.

- **IN 1981,** T. S. Eliot's poetry book *Old Possum's Book of Practical Cats* was transformed into the musical *Cats* by Andrew Lloyd Webber and taken to London's West End – where it ran for 21 years – and New York City's Broadway, in 1982, where it ran for 18 years. It featured the mysterious Mr Mistoffelees, 'the original conjuring cat', who performed extraordinary feats of magic and the most demanding dance routine in the show, including 24 *fouettés en tournant* (pirouettes performed with a whipping movement of one leg to the side).

Did you know...

... that the earliest known cave paintings are in the Chauvet-Pont-d'Arc cave in the Ardèche department of southern France? They date to roughly 30,000–35,000 years ago and hundreds of painted animals from at least 13 species have been catalogued at the site, including predatory animals not usually found in cave art, such as lions, bears and owls.

RAISING THE ALARM

Many animals have superior senses of smell and hearing to humans, and tend to be much more alert – which is why a rabbit could sense that a sleeping woman was falling into a diabetic coma, for example, or a mule out in a stable could realise its owner's house was alight. In such situations, the pet animals in this chapter have proved invaluable at providing their owners with warning, allowing them to escape a dangerous situation, or getting help for someone who has fallen ill, in time to save their lives. Other animals have gone one step further and gone themselves to fetch help – somehow communicating to other humans the need to follow them back to the person in trouble.

LULU

A pot-bellied pig named Lulu outdid a dog in the
heroic stakes in their owner's hour of need…

On 4 August 1998, Jo Ann and Jack Altsman of Beaver Falls,
Pennsylvania, were holidaying at their trailer at Presque
Isle with their pets, an American Eskimo dog named Bear
and a Vietnamese pot-bellied pig called Lulu. The pig had
originally belonged to their daughter Jackie, but when the
couple had agreed to babysit the animal they fell in love with
her and ended up keeping her. Then, she weighed 4 pounds,
but just one year later she had grown to 150 pounds.

That day Jack was out fishing on Lake Erie when his
wife collapsed in the bedroom of the trailer. The 61 year
old had suffered a heart attack within the past 18 months
and recognised what was happening. In an effort to get
attention she threw an alarm clock out of the trailer
window, and Bear began to bark, but there was no one
nearby to hear the alarmed animal and come to her aid.
However, Lulu seemed much more determined to get help.
First she struggled out through a 'doggy door', cutting her
fat belly, then pushed open the gate and carried on down
to the road where she lay down, placing herself firmly in
the flow of oncoming traffic.

When one driver finally pulled up and got out of the
car to inspect the prone pig Lulu got to her feet and led

the man towards the trailer. The man called out, saying he had found a pig in distress and was the owner inside? Hearing the man, Jo Ann called out that she was the one in distress and asked him to call an ambulance. Thanks to Lulu's remarkable actions, paramedics arrived in time to save Jo Ann's life. She was flown to Beaver where she had open-heart surgery – doctors later said that if another 15 minutes had passed it was likely they wouldn't have been able to do anything for her.

After news spread of her heroic deed, Lulu was set on an unstoppable rise to world fame. The story was covered in *The New York Times*, *USA Today* and *People* magazine, and was soon appearing on news channels around the world, from Australia to Italy. Numerous appearances on American chat shows followed, including the *Late Show with David Letterman*, where Lulu met George Clooney. The Hollywood star was himself the owner of a much-loved pot-bellied pig – much to the jealousy of numerous past girlfriends! Clooney was so impressed with Lulu that he got in touch with Oprah Winfrey about the famous pig – and Lulu found herself on a jet being flown in to make a star appearance on the *Oprah* show. Lulu was awarded a gold medal by the American Society for the Prevention of Cruelty to Animals, but more importantly, her owners rewarded her with her favourite treat – a jam doughnut.

LONG-LONG

Although reptiles are cold blooded, that doesn't mean to say they are cold hearted, as one pet snake in China showed…

When Yu Feng discovered an injured snake outside his home in Fushun, in Liaoning, China, he decided to nurse the poor animal back to health. He treated it with herbal remedies for several weeks until he was satisfied it had made a full recovery, then went to release the animal back into the wild at a spot a mile from his home.

To Yu Feng's surprise, the reptile soon turned up again at his house. Puzzled, he released it back into the wild, only for it to return another two times. By this stage, friends and neighbours began telling Yu Feng that perhaps the snake had come back to repay his kindness. And so he kept it as a pet and named it Long-long.

One night Yu Feng and his family were all asleep when he felt something cold on his face, and woke to find that it was Long-long. Bemused but drowsy, Yu Feng turned over and went back to sleep. He woke again to find the snake sinking its teeth into his clothing and whipping the bed with its tail. It then moved to his mother's bed and began to do the same.

Now fully roused, Yu Feng got up and realised he could smell burning. On investigation he discovered his mother's electric blanket had caught fire, and turned it off just before the fire could really take hold. He had no doubt that Long-long had been trying to warn him – and that his debt of a life had finally been repaid.

LOU

When Jolene Solomon's new year started out in a blaze of glory, she was glad to have her loyal mule Lou at her side…

On New Year's Day 2009, Jolene Solomon, aged 63, was eating a relaxed dinner at her home in McMinnville, Tennessee, where she lived alone. Just as she was tucking in, she heard her mule, Lou, braying and kicking up an awful fuss. When the ruckus carried on for some time she got up to investigate.

Stepping outside into the cold air, she saw Lou running from the barn, throwing her head up and whinnying. That's when she looked round at the side of her house to see smoke billowing out and a fire blazing – no wonder Lou had been trying to get her attention! She called 911, but by the time the fire fighters arrived she had watched her home burn to the ground. It looked like she'd be starting off the new year with Lou and not much else.

Jolene's father had bought Lou for her years before to help her and her sister, Blue, around the farm. When her sister had died Jolene had been left alone with Lou, and the mule had become an important companion for her in the months when she struggled to come to terms with her loss.

Jolene would have to stay with family and friends until she could rebuild her home, which her grandfather had built and where she had lived all her life. But, although she had lost so much, she was lucky to have a true friend in Lou, and credited the trusty mule with saving her life.

LULU

Lulu the roo put Skippy to shame with her real-life action heroics in Victoria, Australia...

Lulu, an eastern grey kangaroo, was rescued by the Richards family from her dead mother's pouch, after her mother had been hit by a car and killed. Usually, if a joey has passed the embryonic stage and is already covered in fur when it loses its mother it stands a good chance of survival. They need to be fed lactose-free milk initially, until they can progress to solids, and because of their natural instinct to cuddle up – which comes from living in their mother's pouch – they are very endearing. A kangaroo that has been reared in this way cannot provide for itself immediately if released into the wild, and many end up in wildlife sanctuaries once they grow too large to manage – fully grown kangaroos need at least 2.5 square miles of space to roam around in, and boundary fences at least 6 feet high to contain them. Luckily for Lulu, space wasn't in short supply on the Richards' farm in the Gippsland region of Victoria. Lulu had lived with them for four years when she proved herself to be an invaluable member of the family.

In September 2003, a storm hit the area, and farmer Len Richards, aged 52, was knocked unconscious by a falling tree branch while out checking for storm damage. He could have been there for hours, exposed to the harsh elements, if it wasn't for Lulu dashing back to the house and raising the alarm by making a barking noise to attract the attention of his wife, Lynn. Perturbed by the animal's agitated state and out-of-character behaviour, she went out with her nephew and found her husband lying unconscious in a field.

A year later, Lulu became the first kangaroo to receive a bravery award when she was honoured with the Australian RSPCA's Animal Valor Award, set up in 1998. Speaking of the incident, Len suggested Lulu might also have performed some 'first aid' to help him. He explained how his nephew had found him tipped over to one side, which had saved him from choking on his own vomit, although they would never know for sure whether this had been down to Lulu.

ROBIN

A pet rabbit kicked up a fuss one night when one of its owners fell dangerously ill…

Ed and Darcy Murphy from Port Byron, Illinois, bought Robin the rabbit at a garage sale and gave the animal as a gift to their children Callie, 10, and Dylan, 8. The children loved her and she seemed to settle in well, but one night ten days later she began to behave very strangely.

It was 3 a.m., and Ed was woken by the sound of Robin bouncing around and thumping in her hutch. 'She was going wild,' he said. 'It wasn't like her at all.' Disturbed from his rest, he noticed his pregnant wife Darcy was making some strange snoring sounds, although he didn't pay too much heed to this and tried to settle back down to rest.

But the agitated bunny wouldn't keep still and the noise woke Ed again. This time, when he looked over at his wife, he saw that she had gone an alarming red colour and was hardly breathing. Although Ed didn't know it at the time, Darcy, who was diagnosed with gestational diabetes, had gone into insulin shock. Ed quickly dialled 911 and the paramedics were able to reach Darcy in time and get her the urgent treatment she needed.

A month later, a fully recovered Darcy gave birth to a beautiful baby girl, whom they named Brenna. Darcy was adamant that if Robin hadn't made all that noise to wake her husband that night, she and the baby would have died.

Dr Bonnie Beaver, an animal behaviourist at Texas A&M University, gave a possible explanation for Robin's behaviour: Darcy's diabetes meant she would have been emitting significant ketone odours, which the rabbit would have been able to smell and to which she may have been reacting. She certainly seemed to know something husband Ed didn't, and if it hadn't been for her he would never have got help for his wife and unborn child in time.

The bunny rescues

In Thatcham, Berkshire, Warren Taylor and Kacey Leathers were disturbed one night by their pet rabbit scratching at their bedroom door. They woke to find the house filled with the overwhelming smell of gas. The rabbit was shut into the kitchen at night, and had somehow managed to get out and up the stairs to raise the alarm.

Another pet rabbit saved the lives of an Australian couple in the Macleod area of Melbourne when a fire started in their house while they slept. Michelle Finn and her partner Gerry Keogh were alerted to the danger when the rabbit scratched on their bedroom door to wake them. The fire caused thousands of dollars of damage and the couple had to rebuild their home – and buy a new cage for their hero pet, whose cage had also been destroyed in the blaze.

PEANUT

A father and son were alerted to a fire in their home
by a rather unusual smoke alarm…

One Friday night, Shannon Conwell and his nine-year-old
son Tyler were having a quiet night in, watching a film
together, when they drifted off to sleep on the sofa in their
home in Muncie, Indiana. Also in the room with them
was their pet parrot, Peanut, whom they had bought six
months earlier and who was already an important member
of their little family.

At around 3 a.m., the pair were awakened from their
slumber by Peanut making a call that sounded very much
like a smoke alarm. Smoke was filling the room and fire
was taking hold all around them – the real smoke alarm
was sounding, but it was Peanut's imitative cries that had
roused the pair.

Shannon knew he had to act fast. He grabbed hold of
his son and the bird, and made a run for it out of the
blazing house, and it was lucky they did as their home was
completely ravaged by the flames. Shannon later said that
he'd initially had to fork out quite a large sum to buy the
bird, but that it was best money he had ever spent.

Many types of parrots are able to imitate human speech,
or other sounds, as Peanut did. Parrots do not have vocal
cords – they make sounds by expelling air across the mouth

of the trachea, or windpipe, which in parrots is bifurcated (branches in two). By changing the depth and shape of the trachea they are able to make different sounds.

WILLIE

In Denver, Colorado, a babysitter was glad that an intelligent parrot named Willie was present at the breakfast table…

One morning in November 2008, Megan Howard was babysitting two-year-old Hannah Kuusk, accompanied by Willie, her green-and-white pet Quaker parrot. As the toddler tucked into her breakfast, Megan popped into the bathroom, leaving the child alone with the parrot.

It was then that Megan heard the parrot screeching and squawking and flapping his wings – he kept saying 'Mama baby' over and over again, until she rushed out of the bathroom to see what the matter was. In the kitchen she was shocked to see Hannah choking and turning blue in the face.

Megan quickly performed the Heimlich manoeuvre, saving the little girl's life – but she gave full credit to her avian friend for the important role he played: 'If [Willie hadn't] warned me, I probably wouldn't have come out of the bathroom in time because she was already turning blue – her lips were blue and everything.'

Peanut later received the local Red Cross's Animal Lifesaver Award, which he was awarded during a Breakfast of Champions event, attended by the governor of Colorado and the mayor of Denver.

HENRY

In Scotland, a quiet stroll with the dog turned into a nightmare one dark autumn evening…

One evening in October 2010, father-of-three, David Baird, was out walking his black Labrador, Henry, near his home in Kilbirnie, Ayrshire. The pair were strolling along a cycle path when out of nowhere came a man who launched a vicious and prolonged attack on David, leaving the 40 year old badly beaten on the ground.

Henry didn't waste any time in dashing back home, where he scratched at the front door to get the attention of David's wife, Annabel. Realising something was wrong she watched the dog closely, then followed him as he began to lead her down the cycle path – where she was horrified to find her dear husband lying in a pool of his own blood.

David was admitted to the Royal Alexandra Hospital in Paisley in a very serious condition, and family friends and neighbours were outraged at the unprovoked violence that this poor man – who was a hairdresser and very popular in the local community – had endured. No one knew why anyone would have wanted to attack David, but they all agreed on one thing – his devoted pet Henry had saved his life.

HUGO

Andrew Williams learnt the importance of getting to know your neighbours when disaster struck in his home one night…

Andrew Williams, an engineer, was at home alone in Birch Hill, Berkshire, one weekend while his wife Sarah was away visiting his sister. He was startled to feel a clawing sensation on his face at two in the morning and woke to see a concerned cat staring down at him – especially as he didn't even own one. The furry face belonged to Hugo, his feline neighbour, who had come through the cat-flap downstairs to warn him that his house was on fire.

Acrid, black smoke was billowing through his home, but didn't set off the smoke detector until after he had woken up. Andrew had recently been doing work on his bungalow and had moved the smoke detector to a different spot. Alerted to the danger, he sprang into action, calling the fire brigade and doing what he could to tame the blaze himself. When the emergency services arrived he was treated on site for smoke inhalation, and it was later discovered an electrical fault was to blame for the fire.

Hugo and his brother Harvey had always been regular visitors to the Williams, who were cat lovers and enjoyed petting the neighbour's cats. 'The fire chief said that I had better buy the cat a big piece of fish because he saved my life,' Andrew later said. 'I'm just so thankful to that little fella.'

HOW ALARMING!

In the animal kingdom, many creatures – especially those that are preyed upon – have ways of letting their fellow species' members know when danger is approaching. The most common form is an alarm call, such as those used by birds, small rodents and monkeys. These calls may also be interpreted by animals of other species – for example, many small birds have similar alarm calls and will take cover on hearing each other's. Visual alarm signals are most common in mammals, and are known as 'flagging' – for example, raising the tail to show a white patch of fur underneath as the animal flees from danger. Ants, bees and certain types of fish have a silent method of communicating about a perceived threat – they secrete 'alarm pheromones'.

- **BELDING'S GROUND SQUIRRELS,** native to the western United States, have two types of alarm calls: a churr or trill, a series of five notes given rapidly, to warn of terrestrial predators that pose a less immediate threat, and a high-pitched whistle to signal the presence of an immediate threat, such as an aerial predator.

- **THE EASTERN,** or purple, swamphen, sub-species of which can be found around the world, gives conspicuous tail flicks to signify danger.

- **VERVET MONKEYS,** commonly found in Africa, have different calls to signify snakes, leopards and eagles, with each call eliciting an appropriate response from fellow vervets.

- **WHEN INJURED,** minnows and catfish release pheromones which cause fish nearby to hide in tight schools near the sea bottom.

The sounds animals make:

- **Apes:** gibber
- **Beetles:** drone
- **Bitterns:** boom
- **Deer:** bell
- **Falcons:** chant
- **Grasshoppers:** chirp or pitter
- **Grouse:** drum
- **Hippos:** bray
- **Hyenas:** laugh
- **Tortoises:** grunt

Did you know...

... that male swallows emit false alarm calls during the mating season when females leave their nests, in order to prevent them mating with other male swallows nearby?

CREATURES THAT CARE

Animals are also capable of a quieter form of heroism – by providing companionship, friendship or maternal care, they can make a real difference to the lives of others, whether humans or other animals. They can provide tremendous support to people battling disease or depression, and can help people who struggle socially to communicate better by improving their confidence or making learning fun. The maternal instincts of female animals have led to some touching stories of creatures that are normally considered prey nurturing the offspring of predators, and of wild animals caring for abandoned and vulnerable human babies. This chapter also brings together some tales of truly unlikely friendships between animals of different species.

PIP

A feline friend helped a girl with cancer through the most difficult period of her life and gave her the confidence to hold her head up high…

At the age of 18, Jessica Ford from Salisbury, Wiltshire, was diagnosed with acute lymphoblastic leukaemia. Aside from the gruelling, painful treatments she had to undergo in order to battle the illness, she lost her hair and a lot of weight, and was concerned about how people would react to her changed appearance when she left hospital. She also hated being separated from her pet cat, Pip, and worried that her beloved friend wouldn't recognise her when she came home.

Jessica had been given black-and-white Pip on her sixth birthday as a surprise gift. The two had become firm friends right away and Jessica loved caring for the energetic little kitten from the outset. When she was eventually allowed home for a day visit her worries were proved unfounded: Pip recognised Jessica instantly, running down the garden path to greet her. It was an emotional reunion, and seeing that her cat still recognised her made Jessica realise that even though she didn't look anything like her usual self, she was still the same person inside. 'This realisation helped me a lot when dealing with what was happening in my life,' said Jessica. 'She had such a calming effect on me.' It was just the confidence boost she needed, and she returned to hospital to complete the remaining weeks of her treatment feeling reassured.

When she returned home a few weeks later as an outpatient, she needed Pip's friendship more than ever. Although her loving family were there to offer support, Jessica's friends were busy doing their A-levels, and she missed them. 'Pip was an amazing companion, having the ability to make me laugh on good and bad days. That was so important at the time,' Jessica explained. On some days Jessica would be in too much pain for Pip to sit on her lap, but it was as if the cat had a sixth sense – when Jessica felt really awful Pip wouldn't even attempt to get onto her lap, she would just sit nearby. Jessica's constant companion even waited patiently outside the bathroom while she showered, and would welcome her home with vigorous purring whenever she came back from hospital appointments.

In 2008, Jess nominated Pip for a Rescue Cat Award, organised by the charity Cats Protection, and Pip was selected as a finalist in the Best Friends category. She had certainly been a true best friend to Jessica: 'She truly gave me so much courage and the desire to fight on,' she said.

BARNEY

Parrots are known for their ability to copy human speech, but the roles were reversed when a little boy from Blackburn learned to speak by listening to a macaw...

Dylan Hargreaves, aged four, had severe learning difficulties and had never uttered a single word. His mother Michelle, 33, said he would try to speak but no sound would come out. But that was all to change when she was given three-year-old Barney the parrot as a gift by her partner Rob Hargreaves.

Dylan seemed to be fascinated by the bird, and loved to sit near Barney watching him and listening to the sounds he made. A few months after Barney arrived in their home, Michelle was delighted to hear little Dylan begin to say his first words. It would only be the odd word, but it came out clearly enough for Michelle to understand what her son was saying.

'Every time I gave the bird something to say, Dylan started trying to say the same thing,' she explained. 'I think it's because the bird says things slower than me, which helps Dylan understand.' Thanks to Barney's help, Dylan was soon coming out with words like 'night, night', 'Dad', 'Mum', 'ta', 'hello' and 'bye', and speech experts believed that soon he would be able to move on to two-syllable words. Michelle was convinced that if he did, his first one would be 'Barney', because he loved his feathered friend so much.

Dr Hazel Roddham, a speech therapist from the University of Lancashire agreed that a child with learning difficulties could benefit from a parrot's slow repetition of words, adding: 'If there's some enjoyment, a child is more likely to learn.'

ALBERT

A lonely orphaned elephant got the swing back into his stride thanks to a woolly new friend…

When a six-month-old elephant was orphaned after his mother died in a fall down a cliff, vets at South Africa's Sanbona Wildlife Reserve hoped the baby would be adopted by another elephant cow and stay with the herd. After monitoring the situation for a while they realised this wasn't going to happen – without milk to feed on he would starve – and so they took him to the Shamwari Wildlife Rehabilitation Centre, in the Eastern Cape.

In his new surroundings, Themba, as staff named the new arrival, made an unusual friend – although the first meeting didn't quite go smoothly. Filmmaker and naturalist

Lyndal Davies described how all hell broke loose when a sheep named Albert was placed in the baby elephant's enclosure. 'Themba made a dash for the sheep and chased him around his watering hole,' he said. The sheep hid in a shelter at the far end of the enclosure for 12 hours. Eventually curious, Themba began to approach the shelter and poke his trunk in-between the poles, sniffing at Albert and gently touching his woolly back.

The next day there were developments. Albert seemed to have grown bored with his hideaway and ventured out. Themba was soon at the sheep's side and the two explored the enclosure together – and they made quite a picture, Themba walking along with his trunk resting on Albert's back. From then on, the pair were inseparable.

The centre's wildlife director, Dr Johan Joubert, was pleased with the pairing because the sheep was strong enough to make a good playmate for Themba and could handle the elephant's boisterous play much better than the human staff members. Albert had helped the elephant to gain confidence and to learn affection, too, by becoming a member of his herd.

Any concerns that Themba would end up thinking he was a sheep were overturned when staff noticed Albert had started to copy everything the elephant did – even tucking into thorny acacia bushes with his pachyderm friend. Themba would eventually be released back into the wild when he was ready and, although the pair would be separated, their friendship would have a lasting positive effect on Themba's development.

MZEE

A lonely baby hippo brought an old tortoise out of its shell when circumstances brought the unlikely pair together after the 2004 tsunami hit Kenya…

In December 2004, a frightened, dehydrated one-year-old hippo calf was found near the coast of Kenya by wildlife rangers. It had been separated from its family and swept out to sea when the Boxing Day tsunami flooded the area. Various people tried to rescue the poor creature, including fishermen using their nets. Eventually the rangers took the hippo, christened Owen, in the back of a pick-up truck, to Haller Park in Mombasa. He would have to remain in captivity, since any male hippos that came across him would have killed him instantly.

At the wildlife sanctuary, the abandoned hippo was placed in an enclosure with a 130-year-old Aldabra tortoise named Mzee, which is Swahili for 'old man'. Staff at the sanctuary were surprised when the pair struck up a friendship and became inseparable: they ate together, slept next to each other and spent all day in each other's company, with Owen the hippo following Mzee the tortoise around and licking his face. Mzee would even let Owen put his mouth around his head, showing great trust between the two.

This bizarre pairing became the subject of much media attention, and Owen and Mzee were soon world famous, with director of Haller Park Dr Paula Kahumbu and father and daughter team Craig and Isabella Hatkoff writing a series of books about the animals. Later on, Owen was introduced to an older female hippo named Cleo in a separate enclosure, where he seemed to settle in well with one of his own kind. But the story of resilient Owen and caring Mzee would continue to entertain children for years to come.

CARING CHIMPANZEES

In Nigeria, a disabled boy who was abandoned in the forest found an unusual new adoptive family...

Hunters in the Falgore forest in Nigeria were astonished to find a two-year-old boy living with a family of chimpanzees. They brought him in to the Tudun Maliki Torrey home in the city of Kano, where nursing staff named him Bello.

They believed he had been living with the chimpanzees for eighteen months, having most likely been abandoned. He was both mentally and physically disabled, and had a misshapen forehead, sloping right shoulder and protruding chest. He was thought to be the child of the nomadic Fulani people, who travel great distances across the West African Sahel region and are known to abandon their disabled children.

Most such children would die, but Bello was lucky enough to be found – probably when around six months old, based on the traits he exhibited – by the family of chimps and cared for by a nursing female. The chimpanzees had taught him their ways and behaviour. When he was brought into the centre he dragged his arms along the ground as a chimp would, said Abba Isa Muhammad, the home's child welfare officer, and 'would jump and grunt or squeak like a chimpanzee... He would leap about at night from bed to bed in the dormitory where we put him with the other children... He would smash and throw things.' Even several years later he would clap his hands over his head repeatedly with cupped hands, and make chimpanzee-like noises rather than speaking.

In spite of his unusual behaviour, the staff were fond of him. It was a miracle that the boy had survived, and he certainly wouldn't have done without the care of the chimpanzees.

Alternative childcare

There have been other interesting instances of animals caring for human children:

In Tehran, Iran, in 2001, a 16-month-old toddler went missing from his nomadic parents' tent while they were working in the fields. He was found safe and sound by a search party three days later in a mother bear's den. The female appeared to have cared for the infant during this time. A medical examination showed the toddler was in good health.

When a monkey began taking an interest in their 21-day-old baby, Rohit Khuntia and Kamalini Khuntia, quite naturally, reacted by shooing the creature away – monkeys were usually seen as a menace in the Dhenkanal area of Orissa, India, where they lived. But the little simian continued to visit their house every day to play with their son. Eventually the parents put their fear to one side and allowed the monkey into the family. Before long it was arriving at the house in the morning and taking care of the baby all day. Mother Kamalini described how the monkey would care for her child while she was busy with household chores, looking after him gently, as a mother would, and never causing him the slightest harm.

ANJANA

A female chimpanzee at an animal sanctuary in South Carolina became a surrogate mother to dozens of baby animals…

In 2008, heart-warming images of a chimpanzee mothering two adorable white tiger cubs were published. Mitra and Shiva were born during Hurricane Hannah at The Institute of Greatly Endangered & Rare Species (TIGERS) in South Carolina. Their mother became stressed when their sanctuary flooded and they had to be separated from her for their own safety. Instead, the 21-day-old cubs were adopted and cared for by chimpanzee Anjana under keeper China York's supervision.

The story of Anjana and the cubs gained widespread attention, but it wasn't the first time the chimp had cared for another animal's babies. In China's role as infant animal care giver, she nurtures hundreds of animals when they are born. China looked after Anjana when she was born and the pair had lived side by side ever since. Anjana began to copy her and join in with caring for and raising baby animals. China explained Anjana's responsibilities: 'She gives them a bottle, lies with them and acts as a surrogate mother. She has a close contact and bond, and gives them a nurturing.'

Anjana had raised leopards, lions and orang-utans before she took on the tigers, and Dr Bhagavan, the founder of the institute, praised her for being a wonderful assistant carer.

CROSS-SPECIES CARE

When an animal provides maternal care to another animal that is not its own offspring it is termed allomaternal care – this is commonly seen in species of primates, such as silvered langurs, which will take it in turns to care for another female's child within their group. As seen in the case of Anjana, this maternal behaviour can also cross species boundaries. Here are some more touching stories of alloparents in action...

- **A 12-WEEK-OLD MACAQUE** that had been abandoned by its mother and was close to death was rescued on Neilingding Island, in Goangdong, China, and taken to an animal hospital. Although his health began to improve, he seemed spiritless, until he made friends with a white pigeon. The two became inseparable and the little monkey got a new lease of life thanks to the affection of his feathered friend.

- **A PIG** adopted three tiger cubs whose own mother couldn't feed them and raised them along with her piglets at the Chimelong Safari Park in Guangzhou's Panyu District, China, in 2007.

- **IN 2005** lurcher Geoffrey befriended a baby deer named Mi-Lu at the Knowsley Animal Park in Merseyside after she was rejected by her mother.

- **AT THE SAFARI ZOOLOGICAL PARK** in southeast Kansas, owner Tom Harvey introduced a golden retriever that had just weaned her own pups to a litter of tiger cubs that had been rejected by their own mother. Puppies and tiger cubs take around the same time to develop, and the dog immediately began to lick, clean and feed the cubs.

'Animals are such agreeable friends – they ask no questions, they pass no criticisms.'

George Eliot

'There is in all animals a sense of duty that man condescends to call instinct.'

Robert Brault

NOT SO WILD
AT HEART

The stories in this chapter show that animals we may normally consider to be wild, even dangerous, are capable of showing compassion and behaving in a gentle, helpful way towards those weaker than themselves, whether they be animals or people. Many of these stories happen to be set in zoos, but that does not mean the animals concerned are in any way tame or domesticated. Their heroic deeds call to mind fictional characters such as gentle giant King Kong and the story of the lion that lay down with the lamb.

JAMBO

In August 1986, a visit to the zoo turned into a terrifying
ordeal for the father of one five-year-old boy…

Jambo was a male lowland gorilla born on 17 April 1961
in a zoo in Basel, Switzerland, and was the first gorilla
to be born and raised by his mother in captivity. He was
transferred to Jersey Zoo, founded by Gerald Durrell, on
27 April 1972, where he was to become the star attraction
following a famous incident.

On 31 August 1986, visitors to Jersey Zoo were enjoying
watching Jambo and his family, who had gathered beneath
the wall of the gorilla enclosure. The wall stood at about
chest level to an adult, and so Steve Merritt, who was
visiting the zoo on holiday from the UK with his wife and
two children, lifted his young son Levan onto the wall to
get a better look. He had only turned away for a moment
when Levan toppled forwards and fell into the enclosure.
Hitting the concrete around the edge he lost consciousness
immediately. His father could only look on, terrified, as
Jambo the big silverback approached his little boy. Gorillas
are known for their territorial nature, and he felt sure his
son would be torn to pieces. But then Jambo did something
that changed people's opinions of gorillas forever.

The curious gorilla bent to look down at the boy, then
gently stroked the skin on his back and sniffed his fingers

to get the boy's scent. Turning his back to Levan he squared up to the other inquisitive members of his family, making it clear that they were to come no nearer. Jambo stood guard over the prone figure of the boy in this way for a while longer, until he came round. Members of the crowd shouted out to the boy to stay down and keep quiet and still. But when the frightened Levan began to cry, Jambo became unnerved, and led his family away and into their house at the back of the enclosure.

Levan wasn't out of danger yet. As the gate to the house was closing, a younger male gorilla named Hobbit shot out and ran over towards the boy in a territorial display. At this point two of the keepers jumped down into the enclosure and, armed only with a stick, kept the angry young male at bay. Paramedic Brian Fox was lowered in next, and tended to the injured boy, who had a fractured skull and a broken arm, and was bleeding heavily. Realising that he urgently needed medical attention, Brian called out to be winched up over the enclosure wall with the boy.

The whole episode was captured on home video by Brian Le Lion, and photographed by many other onlookers there that day. Jambo was hailed as a hero – not just for his gentle actions towards the vulnerable boy, but for helping to ease the public's fears about the violent nature of gorillas. Jambo went on to sire many of his own children, several of which are still alive in zoos around the world today. He died on 16 September 1992, and his life story was recounted in the book *Jambo – A Gorilla's Story*, by his keeper Richard Johnstone-Scott. A bronze statue of Jambo was erected inside Jersey Zoo as a tribute to this magnificent beast that helped to change public perception of the species for the better.

Binti

At the Brookfield Zoo in Illinois, a female lowland gorilla named Binti Jua was involved in a similar incident. Binti's own baby, Koola, was just 17 months old on 16 August 1996, when a three-year-old boy fell 24 feet into the gorilla compound of the zoo's Primate World, where Binti and several other adult gorillas were on display.

Binti approached the unconscious boy, and when another female came over she growled at her, keeping her at bay. Next, with Koola already firmly attached to her back, she gently picked up the boy and cradled him in her arms. As touching as this display of maternal instinct was, it was what Binti did next that really showed her intelligence. Onlookers around the enclosure were astonished to see the seven-year-old gorilla carry the boy 60 feet to an access door and set him gently down, where the keepers were able to safely retrieve him. Thanks to Binti's actions the boy was kept safe, and could be taken to hospital where he received medical attention and made a full recovery within four days.

Binti was the niece of Koko, who became world famous in the 1970s for her ability to communicate with humans through the use of American Sign Language, and when Dr Francine 'Penny' Patterson, director of The Gorilla Foundation in Woodside, California, showed footage of the rescue to Koko, the older gorilla seemed impressed. 'She answered "lip", her word for girl, and "good" when asked what Binti had done,' said Penny.

GIMPY

A rescued elephant seal pup in California became the star attraction when she determinedly protected her keeper…

In 1994, an injured elephant seal pup was brought into the Marine Mammal Care Center at Fort MacArthur, California. This hospital for ill, injured and orphaned marine mammals has been caring for rescued California sea lions, northern elephant seals, harbour seals and northern fur seals since 1992.

It was touch and go for the latest admission for a while, but staff at the centre were relieved when she made a full recovery. Christened Gimpy, the pup went on to become a firm favourite among visitors and the local community, and was known as a 'gentle giant' – at over 150 pounds, she was certainly no lightweight.

Volunteer Hugh Ryono was particularly fond of Gimpy, and she would often be at his side while he was at work in the seal enclosure. One day as he was feeding the elephant seal pups, he slipped and fell heavily onto the deck. Gathering his wits he looked up – just in time to see three aggressive pups moving at speed towards him.

Hugh was unable to move and was sure that he was about to be mauled but, just as he reached for his protective board, something extraordinary happened. Gimpy rushed to his side and placed herself firmly between him and the approaching pups, forming a living shield of seal blubber and forcing them back. Later, Hugh expressed his relief that she had been there that day – and had no doubt that she had saved him from a brutal mauling.

ETHIOPIAN LIONS

In Ethiopia, a girl was saved by an unlikely group of rescuers…

In June 2005, *The Guardian* published an intriguing article about a kidnapping in Ethiopia. In the remote southwest of the country, a gang of men had abducted a girl of 12 and were intent on forcing her into a marriage. They held her for a week, beating her repeatedly in order to break her will – not an uncommon occurrence in the area, where young girls are often beaten and then raped.

Fortunately for this one, out of nowhere three lions arrived at the scene and chased the men away. They then stood guard over the terrified, shocked and battered girl, until she was found by the police. Sergeant Wondimu Wendaju of Kefa province said: 'If the lions had not come then it could have been much worse.' The police managed to catch four of the men, but were still on the hunt for the other three.

But what could have caused the lions to act in this protective manner? A wildlife expert from the rural development ministry, Stuart Williams, suggested that the lions could have mistaken the girl's cries for the mewing sound that a lion cub makes, which could explain why they didn't eat her, although why they would rush to her aid and face down seven men remains a mystery.

A wolf among the sheep

Farmer Tomo Spudic from Netretic, Croatia, discovered his pet wolf would protect his sheep from other wolves. He had been given a wolf cub by a friend who thought it was a husky dog. When it turned out to be a wolf he kept it anyway, and named it Vucka. A year later he came up with the idea of using Vucka to protect his sheep, and put him to work in the field with his prize flock. He soon found that Vucka not only left the sheep unscathed, remarkably, but was much more effective at keeping other wolves at bay than a dog would be, since a dog would be prone to attack from wolves. 'Nothing attacks Vucka,' the savvy farmer proudly proclaimed.

NINGNONG

When the Boxing Day tsunami of 2004 hit the coast
of Thailand, one little girl made a lucky escape…

In December 2004, eight-year-old Amber Mason of
Milton Keynes was enjoying the holiday of a lifetime with
her family in the Thai resort of Phuket. Unaware of the
danger building out in the Indian Ocean, where a level 9.3
earthquake had struck, Amber's parents watched as their
daughter took an elephant ride.

Unusually for the four-year-old elephant Ningnong, she
became unsettled and began to act strangely. As the water
suddenly ebbed back from the beach the elephant took off,
running up the beach and taking Amber with her. When the
powerful tsunami swept into land, Ningnong was soon up
to her shoulders in water – but Amber was safe on her back.

Back at home, Amber said how lucky she felt. 'I think
Ningnong thought something was wrong and was trying to get
off the beach.' Amber's parents had fled from the devastating
wave along with so many others and had been terrified that
they had lost their daughter. Sam, Amber's mother, afterwards
had no doubt that the elephant was responsible for saving
her little girl's life: 'If she had been on the beach on her own
or with us on the beach, she would never have lived. The
elephant took the pounding of the wave.'

To show their gratitude, the Mason family pledged to pay £30
a month to Ningnong's owner for the hero elephant's upkeep.

TUK

A polar bear had an unexpected visitor to his enclosure at the zoo one day…

One day in 1983, visitors were gathered around the polar bear enclosure at Stanley Park Zoo in Vancouver, Canada, watching the zoo's most famous resident, a polar bear named Tuk. He wasn't doing much of interest, though – just lying stretched out and having a nap, when suddenly a young man ran up and threw a kitten over the wall and into the polar bear's pool.

Onlookers watched, horrified, as the poor little creature struggled in the freezing water then began to sink. Just then, Tuk rose from his slumber, stretched and slid into the pool after the kitten. A few moments later he surfaced, with the kitten held gently in his teeth, almost as a mother cat carries her young. Emerging from the pool he lay down with the bedraggled animal between his front paws and proceeded to lick it dry. Tuk kept the kitten warm in this way until keepers were able to retrieve it: the gentle bear had saved its life.

In December 1997 Tuk died at the age of 36 – he was the last remaining animal at the zoo, having stayed behind due to his old age, and with his passing the zoo closed its doors for the last time.

SHOOTER

An unusual candidate stepped up for lifeguard duty
at a zoo in Idaho...

One day in June 2011, volunteer Joy Fox was at work at
Pocatello Zoo, Idaho, when she noticed one of the residents
behaving rather strangely. Shooter, a four-year-old male
elk, had come up to the drinking tank in his enclosure and
begun pawing at the water and then circling the tank. Not
sure what to make of this, Joy grabbed her camera and
captured a remarkable incident.

After continuing in this way for about fifteen minutes,
Shooter managed to manoeuvre his antlers so that he could
reach into the water and pull something out in his mouth.
Joy was amazed to see the elk lift a yellow-bellied marmot
clear of the tank and then put it on the ground, where he
nuzzled and pawed at it gently with his hoof. Eventually
the soggy – and no doubt somewhat bewildered – marmot
came round and scampered off.

Zoo officials were stunned by the events, and zoo
superintendent Scott Ransom said: 'It's hard to tell if he
was helping or didn't like the animal swimming in his
water.' But colleague Kate O'Conner felt sure the elk had
acted with intent to save the rodent: '... we saw it down
on the ground and he was nudging it and it was moving.
He deliberately took that animal out of the tank,' she said.

Wild marmots are a common sight on zoo grounds, and following the incident staff decided to install safety ramps on the larger drinking tanks to avoid the same thing happening again. A week after the story was published in the local media, Google had tracked nearly two million hits on the story and Ransom had received calls for information on the rescue from newsrooms throughout the United States and abroad.

THEIR LIVES DEPEND ON US

Sadly, many wild animals are under threat due to poaching, people encroaching on their habitat and the effect of man's activities on the global climate. As magnificent and powerful as animals such as gorillas and polar bears are, they cannot survive indefinitely in the face of such an onslaught. If future generations are to enjoy more stories of animals' exploits, such as those featured in this book, and their benefit to the human race, then we must strive to protect them. Even the tiniest creature is precious, and while a funny-looking bug or a worm might not have the ability to save a human from drowning or boost the morale of soldiers in war, it still deserves its place in the world and contributes invaluably to the ecosystem in which it exists. There are many organisations and charities that strive to safeguard the world's wildlife, each with its own inspiring story to tell – a few of which have been mentioned below...

- **DURRELL WILDLIFE CONSERVATION TRUST** – was founded by world-famous author and naturalist Gerald Durrell, who had a vision of creating a 'stationary ark' where endangered animals could be bred in captivity. Based at its headquarters in Jersey it works with partners all over the world and has made a significant contribution to safeguarding more than 30 species. Part of that mission has been to create 'Durrell's Army': wildlife conservationists from all over the world who are trained in Jersey and go back to their own countries to save animals for themselves.

- **WWF** – the world's leading independent environmental organisation, the World Wildlife Fund works to protect not only endangered species but also their environments. It additionally works with governments to tackle climate change. Set up in Switzerland in 1961, it has grown into a truly global organisation.

- **BORN FREE** – after starring in *Born Free*, the film that told the true story of Joy and George Adamson's fight to return Elsa the lioness to the wild, Virginia McKenna decided to set up an organisation with the vision of keeping wildlife in the wild. It works with animals in their natural habitat, encouraging people to live peacefully alongside wildlife, and rescues animals that have been confined in inhumane conditions. Its ultimate aim is for zoos to be phased out and for all wild animals to be able to live in their natural environments.

Did you know...

... that fishing is central to the livelihood and food security of 200 million people, especially in the developing world, and that one in five people on the planet depends on fish as their primary source of protein? That makes over-fishing a major threat not just to the species of fish affected but to the lives of millions of people.

'All over the world the wildlife that I write about is in grave danger. It is being exterminated by what we call the progress of civilization.'

Gerald Durrell

'Who will raise their voice when mine is carried away on the wind?'

George Adamson

USEFUL RESOURCES AND INFORMATION

Born Free Foundation

The Born Free Foundation is an international wildlife charity, devoted to compassionate conservation and animal welfare. It takes action worldwide to protect threatened species and stop individual animal suffering. Born Free believes wildlife belongs in the wild and works to phase out zoos. It rescues animals from lives of misery in tiny cages and gives them lifetime care, and also works with local communities to help people and wildlife live together without conflict. Every year, Born Free helps hundreds of thousands of animals worldwide.
Telephone: 01403 240 170
Email: info@bornfree.org.uk
Website: www.bornfree.org.uk

Cats Protection

Cats Protection is the UK's leading cat welfare charity and helps over 193,000 unwanted cats and kittens through a national network of 253 volunteer-run branches and 29 adoption centres. The charity's vision is a world where every cat is treated with kindness and an understanding of its needs.
National helpline: 03000 121212
Website: www.cats.org.uk
Email: helpline@cats.org.uk

Dogs for the Disabled

Dogs for the Disabled is a pioneering charity that trains dogs to carry out a range of practical tasks to assist disabled children and adults in order to achieve greater independence. The charity also provides a programme to support parents of autistic children. Since 1988 the charity has trained more than 400 partnerships.
Telephone: 01295 252 600
Website: www.dogsforthedisabled.org

Durrell Wildlife Conservation Trust

Durrell Wildlife Conservation Trust is an international charity working globally towards our mission of saving species from extinction. Committed to conserving the diversity and integrity of the life on earth, Durrell has developed a worldwide reputation for its pioneering conservation techniques.
Telephone: 01534 860 000
Website: www.durrell.org

Enchanted Forest Wildlife Sanctuary

The Enchanted Forest Wildlife Sanctuary, Inc., is a small non-profit organisation founded by Eve and Norman Fertig, licensed wildlife rehabilitators, with New York State Department of Environmental Conservation and the United States Fish & Wildlife Service of the Department of the Interior. Eve and Norman personally care for sick, injured, orphaned and distressed wildlife and prepare them for survival and return to the wild.
Telephone: (+1) 716-681-5918
Email: IAMSPOCK@aol.com
Website: www.nfwhc.org/enchanted.htm

Gentle Carousel Miniature Therapy Horses

Gentle Carousel Therapy Horses is a non-profit charity based in Florida. Their 24 tiny therapy horses bring their special love where it is needed most, working inside hospitals, assisted living programs, hospice programs, programs for Alzheimer patients and with adults and children with disabilities. They also work with foster children and at risk and abused children.
Telephone: (+1) 352-226-9009
Email: MiniHorseTherapy@att.net
Website: www.horse-therapy.org

Imperial War Museum

The Imperial War Museum is unique in its coverage of conflicts, especially those involving Britain and the Commonwealth, from the First World War to the present day. It seeks to provide for, and to encourage, the study and understanding of the history of modern war and 'wartime experience'.

Telephone: 02074 165 320
Email: mail@iwm.org.uk
Website: www.iwm.org.uk

Marine Mammal Care Center

The Marine Mammal Care Center in Fort MacArthur, California, is a hospital for ill, injured and orphaned marine mammals. Their primary work is the treatment and release of rescued California sea lions, northern elephant seals, harbor seals and northern fur seals. This work is authorized by the National Marine Fisheries Service (NMFS).

Telephone: (+1) 310-548-5677
Email: info@marinemammalcare.org
Website: www.marinemammalcare.org

National Search and Rescue Dog Association (NSARDA)

The NSARDA is an umbrella organisation for Air Scenting Search Dogs in the UK. Its members are the Search & Rescue Dog Associations (SARDA), which are located throughout the UK. Each SARDA is a voluntary organisation responsible for the training and deployment of air-scenting search and rescue dogs to search for missing persons in the mountains and high moorlands of Britain as well as the lowland, rural and urban areas.
Website: www.nsarda.org.uk

Pets As Therapy (PAT)

Pets As Therapy, or PAT for short, is a registered charity that provides therapeutic visits to all kinds of health establishments by volunteers with their own friendly, temperament-tested cats and dogs. The charity operates on the premise that, while research on the proven medical benefits of the company of cats and dogs is still developing, there is no doubt that many sick or ailing individuals appreciate the warm company of a furry feline or canine companion. All PAT dogs and cats are certified by accredited assessors and given the necessary checks by qualified vets before being permitted to make any visits.
Telephone: 01844 345 445
Website: www.petsastherapy.org

PDSA

Founded by Maria Dickin in 1917, the People's Dispensary for Sick Animals provides more than 2 million free veterinary treatments each year, funded completely by public support.
Telephone: 0800 731 2502
Website: www.pdsa.org.uk

Rocky Ridge Refuge

Rocky Ridge Refuge in Gassville, Arkansas, is run by Janice Wolf, who rescues and cares for a variety of creatures, both exotic and the more common.
Website: http://rockyridgerefuge.com

The Guide Dogs for the Blind Association (UK)

This organisation exists to provide guide dogs and other mobility services that increase the independence and dignity of blind and partially sighted people. They campaign for improved rehabilitation services and unhindered access for those who are blind or partially sighted.
Telephone: 01189 835 555
Website: www.guidedogs.org.uk

Urban Search & Rescue dogs (USAR)

The team was officially formed in July 2001 to give the UK Fire & Rescue Service a search and rescue dog team capability to respond to UK emergencies as well as overseas disasters. The teams are made up of fire fighters from individual brigades throughout the UK, who are on 24-hour standby, 365 days a year.
Website: www.ukfssartdogteams.org.uk

WWF

WWF-UK is the UK arm of the WWF Network, the world's leading environmental organisation founded in 1961 and now active in over 100 countries. Their 300-strong staff work with businesses and communities in the UK and around the world to protect not only endangered species but also their environments. It additionally works with governments to tackle climate change.
Telephone: 01483 426 444
Website: http://www.wwf.org.uk

DO YOU HAVE YOUR OWN ANIMAL HERO STORY?

If you have your own stories of amazing animal bravery, we'd love to read them. Please send them to us at the address below and we'll include the best in any new edition of this book:

Ben Holt
c/o Summersdale Publishers Ltd
46 West Street
Chichester
West Sussex
PO19 1RP

DOG
HEROES

TRUE STORIES OF CANINE COURAGE

BEN HOLT

DOG HEROES
True Stories of Canine Courage

Ben Holt

ISBN: 978-1-84024-767-1   £7.99   Paperback

🐾 Swansea Jack, the Labrador that rescued 27 people from drowning

🐾 Max, the collie cross that warned his owner that she had breast cancer

🐾 Shadow, the Rottweiler that saved three young children from a pair of hungry wolves

These are just a few of the inspiring true stories in this collection of dog tales from around the world. Included are some astonishing first-hand accounts by people who have witnessed quick-thinking and resourceful dogs in action.

From trained lifeguard dogs and guide dogs to loyal family pets and unnamed strays, each of these courageous canines has shown true heroism – sometimes in the most surprising of ways. Heart-melting, dramatic and often deeply moving, *Dog Heroes* proves why dogs can save and change lives, and are truly our best friends.

WONDER CATS

TRUE STORIES OF EXTRAORDINARY FELINES

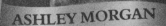

ASHLEY MORGAN

WONDER CATS
True Stories of Extraordinary Felines

Ashley Morgan

ISBN: 978-1-84953-042-2   £7.99   Paperback

🐾 Unsinkable Sam, the courageous feline that escaped three shipwrecks during World War Two

🐾 Mia, a stray that survived an epic journey from Hungary to the UK on a lorry and gave birth to two kittens along the way

🐾 Fred, the undercover cat that assisted the NYPD in the arrest of a suspect posing as a vet

How much do you really know about the abilities of domestic cats? Whether they're using up one of their nine lives to make a lucky escape or demonstrating unusual talents, cats are full of surprises.

From loyal moggies to courageous life-saving kitty heroes and extraordinary survivors, *Wonder Cats* contains heart-warming and moving tales that demonstrate the astonishing powers of remarkable felines.

THE DOG LOVER'S COMPANION

Vicky Barkes

ISBN: 978-1-84953-159-7   £9.99   Hardback

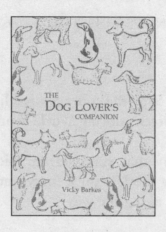

*There is no psychiatrist in the world
like a puppy licking your face.*

Ben Williams

*Dogs are our link to paradise. They don't
know evil or jealousy or discontent.*

Milan Kundera

This beautifully illustrated miscellany is a must-have for any dog lover. With quotations, stories, tips and trivia, along with classic poems by Kipling and Barrett Browning, this delightful collection is as warm as a faithful furry companion at your feet.